The Easy Guide to Solar Electric

by Adi Pieper

ADI Solar
Santa Fe, New Mexico
1999

Farmers Branch Manske Library
13613 Webb Chapel
Farmers Branch, TX 75234-3756

The Easy Guide to Solar Electric
Published by
ADI Solar
290 Arroyo Salado
Santa Fe, New Mexico 87505
www.adisolar.com

Copyright ©1999 by ADI Solar
All rights reserved. No part of this book, including cover
design, interior design, or logos, may be reproduced or trans-
mitted in any form, by any means (electronic, photocopying,
recording or otherwise) without the prior written permission of
the publisher.

Library of Congress Catalog Card No. 99-66143

ISBN: 0-9671891-0-1

Printed in the United States
10 9 8 7 6 5 4 3 2 1

Distributed worldwide by ADI Solar

Book Design and Editing by Barbara H. Johnson
Cover Design and Illustration by Mark De Francis
Chapter Illustrations by Adi Pieper

Farmers Branch Manske Library
13613 Webb Chapel
Farmers Branch, TX 75234-3756

ACKNOWLEDGEMENTS

Many years of work went into this book. It started out as a small 100 page booklet under a different title. With a lot of feedback by the Chelsea Green publishing house, it grew almost double in size; but it eventually fell through the cracks. However, the manuscript found a lot of support with those who especially liked the way it was written, since it approached the subject from a unique angle.

My thanks go to those who initiated its process, like Natasha, and those who supported me along the way, like Bettina, Barbara, and Bob D.—in particular to my editor, Barbara Johnson, who had to put up with a German-thinking brain trying to write English.

Special thanks also to Mark De Francis who understood my strange and sometimes biting sense of humor and designed the cover. Time did not permit him to do the rest of the drawings, which I did myself.

Also special thanks to Windy Dankoff from Dankoff Solar who supported my efforts and "kinda liked the humor in the book."

I also want to acknowledge all the negative critics who did not like the many "wisecracks" in this book. To them, I want to say: "Some lightening up is required!"

iii

TABLE OF CONTENTS

The Easy Guide to Solar Electric

Introduction
WHY WOULD ANYONE WANT TO LIVE IN A SOLAR HOME?

This book is designed to answer some of your questions about solar energy, such as, why would one choose solar energy over conventional energy? We will be talking not only about the "why" of things but also and foremost about the "how."

But first things first: Why? Recall the famous last words of the notorious Timothy Leary: "Why? . . Why not!" Well, between those dots there is a lot one can say.

My own example will show you one way people get introduced to solar power—which is through the door of money, or rather the lack of it. Solar energy is cheap and gets cheaper each year as solar equipment enters the age of mass production. It is no longer just for idealists in the desert somewhere who try by all means to survive the alternative way.

1

Solar hot water and the passive solar building style have entered the mainstream in the sunbelt states, and now getting your electricity from the sun seems to be more and more a natural thing to do.

As property prices get higher near urban areas, people start looking for land in remote areas. However, utility companies charge for hookups, which in many cases would more than make up for the money one could save on a remote lot. In those cases, a photovoltaic system is the only answer and not just an alternative.

An interesting example of solving these remote disadvantages can be found in Australia where the government pays for the main part of a solar system if the utility hookup costs more than a set amount (last I've heard—around $8,000.)

But government

support on the left, Freemen on the right, buying and living remote, or just following common sense while the sun is shining, choosing solar power over conventional or

grid power should not be a major nightmare.

Another advantage of solar electricity is that it is environmentally clean, minimizing the electromagnetic fields that surround you every day and whose effects on the human body and psyche we know very little about. So I will dedicate **Chapter Eleven** (p. 162) just to this subject.

2

The main problem I encountered, working in the field for over seven years, is that people avoid something they do not understand. Not that they understand conventional electricity, but everybody has it and so it becomes a normal thing to live with. But, when interested people start to inquire about solar electric, they often find themselves confronted with such an overwhelming flood of information, indecipherable terms, and questionable promises that, in the end, they either give up altogether or give up control of the design, installation, and maintenance to someone else, never really knowing why things work the way they do and what the limitations of their system are.

Time and again, I ran into people who had been living in solar homes for years who would ask me, "Can you please explain to me, in simple language, how my system works?" There are more women living by themselves in solar homes than one might expect, and, even though they have managed to live with their system for some time, many have no clue about the gadgets and switches they found installed in their homes. Nobody had bothered to explain these things or, if someone tried, they could not follow their "expert" language.

So I decided to take on the task of translating the solar-techno-language into simpler terms. As you will discover in the chapters to follow, I myself had to go through a bit of a learning process and still, today, when I read the description of a new item, I shake my head, go to the phone, call the technician of the company who sells the product, and yell at him. Not that it changes anything as far as understanding their lingo is concerned, but after hanging up or being hung up on, I can start all over again with a relaxed mind.

Responses by solar

experts to early drafts of this book only confirmed what my customers complained about. The "experts" started to call me names, called my book "politically incorrect," said it had too many wise-cracks and not enough depth. In all, I got the strange feeling that they like to be rather dry, humorless, and not understood. "This is nonfiction," I heard them say, "this should be educa-tion and not entertainment." I can't rid myself of the sus-picion that they do not un-derstand that a dry and seri-ous matter can often best be taught through light enter-tainment. (I did not get a single complaint from any-body but the professionals.) So, dear reader, brace your-self for a roller-coaster ride powered by solar electric.

The intention of this book is not to make you an expert technician or a solar engineer, but to make you expert enough so that you can ask appropriate ques-tions when a solar salesper-son runs you, with the speed of light, through several thousand dollars worth of equipment and you only understand the words: "Sign here!"

This book will also enable you to understand the system you have been living with for some time. I will introduce you to the basic terms of electricity. At times I use simple formulas as a back up, but never as a sub-stitute for language. You do not have to read them, you may skip them and still get the picture.

If you plan on in-stalling your own system, you will find enough infor-mation to get a basic under-standing of how things work. However, I consider it a task big enough to dedicate a whole book to. If you have some electrical knowledge and are familiar with conven-tional wiring techniques, this book may give you the addi-tional information you need. If you are a total beginner, you may want to reconsider and do some additional studying. **Never underestimate the power of DC current.** Poor workmanship, too thin wires, or reverse polarities can ruin expensive equipment in a spark of a moment.

It is important to understand that your workmanship has to be exceptional and there are no shortcuts possible. After all, you may be playing with fire, so,

when in doubt, ask for the supervision of an expert.

When you leaf through a book describing a technical subject, just look at the graphs, diagrams, and tables, and you will see just how much you will understand. You will find very little of these in this book. (I always liked books which explain a complex process and, in the middle of the sentence, refer to tables XXIIa and XXIIb. After which you stop reading and start looking for these tables, only to find that, once you find them 10 pages ahead, you have lost the thread of what you were reading.)

I have tried to introduce the subject with humor and picturesque examples. If these examples don't always represent real life scenarios, I hope you will forgive me. At times my humor may be a bit edgy, which just reflects the amount of frustration I have encountered on my own solar path. To all those "experts" I can say, I'd rather read through a book filled with humor, however strange it may be, than bore myself to death with slow statistics and unreadable graphs. (Being of German origin, I do not claim to know much about humor. After all, the smallest book ever written is **500 Years of German Humor.**) I even would, on occasion, take into account a bit of political incorrectness. (Now I have to find a way to make this a "hidden file" which only shows up in print, because my editor will surely "blue pencil" it all.)

It is probably advisable to start reading this book at the beginning in order to follow the fine thread of examples that winds itself from chapter to chapter. If you are a solar expert already, this book may help you understand why, at times, people look at you in a certain strange and unexplainable way.

The reason I use formulas at times is to demonstrate that there is a balance between power available and power required. And in a solar home, you always have to think about this relationship because your lifestyle changes with this balance.

Usually you have two known values like volts or watts and, by multiplying or dividing those values, you find out about a third value, like amps. If you change any value in a formula, all others will change, too, because they are all trade-offs. If you take away from one side, the other side will increase. The pivotal point of each equation is the "="! Once you cross it, you move into the other direction. If you multiplied on one side, you will

have to divide on the other. Those are principles of math. With electricity, you always have one value fixed. A motor, for example, delivers a fixed amount of horsepower or watts. In order to run this motor at its fixed value, you can trade amps for volts. More volts require less amps, etc. Once you understand this principle, the following chapters will be easy to comprehend. But—read for yourself. We still have a way to go before solar electricity will make some sense.

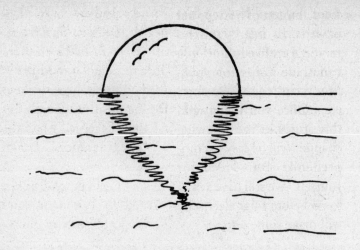

Chapter One
THE BEGINNING

At some point in my life, I decided that enough was enough. The urban California lifestyle was becoming too costly. The amount of energy and time I had to put into just making ends meet was outrageous. Every month, I paid over $800 in rent, plus utilities—money literally going down the drain. I thought that, if at least I could pay this money into a mortgage, it would not all be lost.

So I ventured out to look for quiet, to be self-employed, and to become my own landlord. But, I soon found out that my savings were barely enough to cover the closing costs for a small house in this part of the world. There would be nothing left for the required down payment. A friend in his wisdom told me that the secret lay in buying land and building my own home at my own pace. Following his

8

advice, I started looking for vacant land. But there was very little that I could afford in southern California. So I went further east and further east until, one day, I ended

up in New Mexico, where land was cheaper.

Still it was quite a task, since my budget was quite restricted. But one day, I found land. I felt like the explorer of a new world. The land was cheap. It had only two flaws: no electricity nearby, and no water underground.

"Have you heard of solar electric?" my wise friend asked me. Sure, I had heard about it. Spaceships use it, and I had seen those little rectangular modules in mail order catalogs, the ones for the great outdoors. But then I realized I had the "great outdoors," lots of it.

"Let's investigate the matter," I thought, and I bought books on photovoltaic systems. I started reading and reading, looked into tables about regional solar gain, declination angles of the sun, time zones, compass variations, and demand factors until my brain was buzzing like a generator. I understood very little of what I read but I gathered enough to know where north and south was and that you needed a roof on which to place your solar array. With the help of my wife, my brother-in-law, and a noisy generator, I started to frame my house. When it came to the point where I was ready for the sun to help, I decided to take the matter to a knowledgeable person.

There was a shop in town which sold solar equipment. One morning at 9 I went there. The shop was closed. "We open at ten

o'clock," said the sign. At 10 I waited in front of the store. At 10:30 I still waited. At 10:45 a blue VW Bus drove up. It looked somewhat like an old-fashioned spaceship which had had trouble getting back into space again and had decided to battle it out on the horizontal plane. All over the roof were mounted rectangular panels, which reminded me of those I had seen in the mail order catalogs, but bigger. Wires were hanging down, were wound around outside mirrors, and fed through half-open windows.

The pilot of this vehicle looked very similar. Hair and beard hung down to his waist and his clothing reminded me of the last episode of *Star Trek*.

"My name is Alfred," he said. Well, things went the way they started, very alien.

"Have you done your load calculation?" he asked me.

"I think I know what I want," I said. "I need electricity in my future house."

"So, you know nothing." He pulled out a work sheet and we filled out all the lines on it.

"Do you want AC or DC?" he asked. Puzzled, I looked at him. "Do I have a third choice?"

He continued to ask questions I had never thought of. How many lights, outlets, pumps, radios, stereo systems, fridges?

"Do you watch TV?" he asked. I had to admit it.

"I watch **Star Trek**." I said. That was acceptable.

After we had filled in all the lines of the form he said, "Now you add up lines one through 12, multiply line 13 by 1.5, divide line 14 by line 13, multiply everything by seven, go back to line 12 and subtract all the time you watch soap operas on TV, then take the square root of 365 and hit it with a two by four. But don't use a pressure-treated one."

At least that's what it sounded like to me. I left the store dizzy, as if I just had returned from a space flight. My pick-up truck was loaded with all kinds of boxes, long ones, short ones, square ones, and yellow ones, and I was several thousand dollars poorer. It was all written down on paper, every step, every function of every item—all I had to do was get a degree in electrical engineering. Which I didn't.

Those were the be-ginning times of home solar electricity, and fortunately we have come a long way since then. Hippy Alfred is still notoriously late in opening his store, but a lot of people have learned to get up earlier and do without him. Systems have advanced to a degree that does not require deductions for watching soap operas. Installation procedures are much simpler and can be done without a degree in engineering. Sophisticated computer equipment, as well as power tools, water pumps, and boilers, can be run on solar power. There is no reason why one has to live in urban areas just because there is no grid power available in rural areas.

Solar houses are electrically cleaner. When you do not use any electricity, nothing is switched on and buzzing inside your walls at 60 cycles per minute, creating electromagnetic fields.

The trend away from cities has hit one more time.

But now it is not just a dream for "drop outs." The age of computer modems and telephone lines makes it possible

for a lot of people to live far away from civilization and still be connected, to run a business from their home. The initial cost of a solar system may be the same as many years of electrical bills, but you are "off the grid." You are not at the utility company's mercy. You do not have any power lines running through your property. You use a natural resource available in abundance. You will change your lifestyle. But almost everybody who has done it agrees that the change was for the better.

Chapter Two
PLANNING AHEAD

Before we start diving into the juicy stuff, a few thoughts about general planning for a solar home, whether you plan to build one or to buy one. When you start thinking about "off-the-grid-living," you already have made a basic decision (be it a conscious choice or by default): you want to be independent and you want to save planetary energy by using renewable resources for your energy needs. Therefore, building or buying a house that is, energy wise, totally inefficient would certainly defeat the purpose. Even if you had all the money you needed to make up for those inefficiencies, you would still use up non-renewable resources in manufacturing other equipment, like heaters, photovoltaic panels, etc., not to mention the additional gas or other fuel needed to compensate for your losses.

To make it simple, a solar house has to be as efficient as it can be for things to work in a balanced manner. It should be built in a passive solar style. It should be sealed and insulated top, bottom, and sides.

R-value of 2. (R-value is the thermal resistance of any housing component, or, in other words, the insulating capacity of the material. The higher the R-value, the higher the insulating capacity.) There are new windows on the market that range as high as R-9. So, if at all possible, don't save when it comes to buying windows.

But even if you can't afford the top-of-the-line

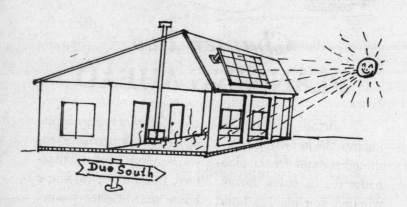

Due South

A passive solar home has three times as many windows as a "normal" home. The heat loss through windows can be enormous. Older style windows, even when double paned, have an

windows, there are other considerations: to seal off windows during nights or during the cold seasons. For example, "thermal curtains" are thermal shields camouflaged with curtains. The outer material is of high R-value, either in the form of shutters or bubble insulation which is flexible, and the inner material is of normal curtain material of your choice and taste. (Also see, *The Passive Solar House* by James Kachadorian.)

The latest trend in heating systems is radiant floor heating. This opens up a whole new world as far as your energy needs are concerned, and it is in effect a very viable solution for passive solar houses which may interest even those people who always said, "Well, solar starts with an 'S,' like in sweaters." But we will talk more about this heating technique when we talk about DC power. (See *Chapter Six "E,"* *The DC System*, p. 62.)

Not every plan for a passive solar design necessarily means that it will work in the real world and in the distant future. An example of this is a solar development near Santa Fe, New Mexico,

started during the initial boom of solar use under the Carter Administration. The local power company supported the passive solar design by supplying cheap electricity for electrical baseboard heaters as a back-up heat source. Every house had two breaker boxes and two meters to supply the designated energy for the heaters only.

Then, political power changed. Policies also changed and solar was out. The price of electricity for those back-up heaters changed as well. On top of it, many contractors with very little knowledge jumped into the initial passive solar market and built houses that did not meet any standards at all: roughly aligned to the south, but often just following other main features like roads, arroyos, etc.; sometimes with single glass windows, often too many of them; hardly any solar mass to collect energy from the sun; and often executed with very poor workmanship so that it soon became apparent that many of those back-up heaters were needed as a fulltime heat source in the winter time at normal electrical rates. The place was called Eldorado. The questions remains, for whom?

To avoid surprises like this, make sure you know at least as much about the subject as your solar architect, if not more.

In the planning stage of your future home, there are several considerations, not only about your present needs but also about your future needs. Most of the older solar generation is deeply rooted in the past, which means their systems are old and for the most part obsolete. The future of solar power however shines as bright as the sun.

Technical development in this field moves as fast as in the computer world (probably because there is a wedding taking place between the two). Whenever I have not been in my supplier's store for a week or two, I hardly recognize the products on his shelf.

So, when you plan your house, you plan for the future. You want to consider designing a system that can constantly grow and change as your needs change and as new technology becomes available. In the chapter on how to design a system, we will go more into more de-

tail. (See **Chapter Seven**, p. 88.)

There are wonderful books written about how to plan and build a solar home and I am not out to compete with those authors. I only want to pull in some of the concepts of a solar house and show how they tie in with the one part we are talking about here, the solar electric.

If you have a lonely log cabin somewhere in the forest, and all you need are some additional lights, you might as well disregard this chapter. But if you want to plan a 2,000 square foot independent family dwelling, you need to remember that an architect's one mistake, placing the living room on the wrong side of the house, for instance, may increase your heating bill, which in effect may create a need for a bigger photovoltaic system, because your radiant floor heater is running for most of the day. Remember, you spend most of the day in your living room and most

of the night in your bedroom (hopefully under thick blankets).

If you are a gourmet cook and want to watch **The Frugal Gourmet** on your

kitchen TV, you not only need to plan for a TV receptacle in your kitchen but also for an exhaust fan. However, you could instead place an operable window over your stove and crack it open while cooking. (This would not substitute for the TV receptacle.) The rising heat with all its gourmet smells will exhaust through this window without using any energy. Not considering the marvelous view you may have, which may add additional spice to your cooking and take care of your TV desire.

My idea of a solar home is one that provides you with all the possibilities that life has to offer, but only as a potential. Which means, your walls don't have to buzz with electrical energy and your circuits don't have to hum just because you want to flush your toilet. (There are toilets that use DC-power to flush.)

But certainly it can't hurt to lay out your wiring in such a way that you have covered future eventualities. Planning especially involves your personal habits. (Not those of your architect!) If you are a morning person, have your living and/or bed-room face towards the south-east so they warm up early with the morning sun. If you have spectacular views to the north, and you must have your living room facing this direction, try to connect your north-facing living room with a south-facing sun room which is situated slightly lower than the living room. The warm air will rise into your living room and drastically cut your heating bill.

For every situation, there are a number of differ-ent solutions with different advantages. Compromises may have to be made. But before you make the final

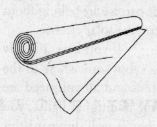

copies of your plans, have a solar electrician, a solar plumber, and a solar heating expert look them over and give them their blessings. You will be surprised at how many changes those people will suggest.

But now let's look into what we want to talk about in this book: electric-ity. And let's start with some basics, the search for the an-swer to "What is Electricity?"

Chapter Three
WHAT IS ELECTRICITY?
The Lovers

In searching for a definition of electricity I opened **Webster's Encyclopedic Dictionary**, and I found, among other definitions which I had a hard time understanding, the following: "Intense emotional excitement." To that I could relate." "When a man loves a woman. . ." was playing in the background on the local radio station.

In a relationship, you certainly go through a spectrum of emotional excitement. Let's just stick with the good parts of it: the attraction. When two people are attracted to each other, they can feel the electricity between them. After that is felt, almost every effort they make in their lives is directed towards coming together. Sometimes they have to overcome a lot of obstacles (*resistance*) and need a lot of push-

19

ing (*volts*) to make speedy progress (*amps*). This may result in a powerful wedding (*watts*). If this is too oversimplified for your taste, here is another version on the same theme:

The electron theory states that all matter is made

of electricity. Usually, an atom has a fixed number of electrons which cannot be moved away from the atom. Electrical phenomena occur when some of these electrons are moved away due to attraction from another atom or due to disturbance of the electrical balance within the atom.

I leave it up to you which one you like better. But does any of the above really explain what electricity is? I tried to investigate. I asked an electrician. He said he was only concerned with installing electrical components and then turning them on. That turned me off. I asked an electrical engineer. He expressed his frustration because he had once searched for the same answer. But all he learned during his studies was how to calculate electrical occurrences. He suggested that I contact a physicist. The physicist in essence told me that he was not even sure whether any electrical phenomena exist in this universe. He added that it probably depended on who was observing them. He suggested that I contact a philosopher. I will spare you the answer of the philosopher; it was like looking for a black cat in a dark room which had just passed through a black hole. In the end he suggested that I contact a priest. I didn't follow up on his suggestion because I knew what the priest would have suggested. Yes, you are right, he would have told me to ask an electrician!

Now that we have thoroughly investigated the question of what electricity is, we can conclude that a good love life will get us closest to experiencing its true meaning. In connection with electricity, we usually are confronted with several electrical terms: *Volts, Amperes (Amps), Watts*, and *Resistance*. We have already encountered them in our little love story. In our daily life we do not need to concern ourselves too much with those terms, except when buying a new light bulb. And then our main concern is whether we want more light or less.

To shed some light into the darkness of these terms, here is a little story about a dog sled in Alaska.

It was a cold and stormy night. The first snow of the season had fallen. A big sled, loaded with emergency gear weighing 1,200 pounds (*watts*) needed to be transported as soon as possible to the next village which was 10 miles away on the other side of the Muchundra Lake. Tom raised dogs. He had 12 dogs (*volts*) available to move the sled. Each was strong enough to move 100 pounds (*amps*). He calculated that 12 dogs times 100 pounds each could indeed move the 1,200 pounds. (12 **V** x 100 **A** = 1,200 **W**).

However, this morning two of his dogs had fought, hurting each other's paws. It would take days before they would be fit again. So he did not have enough dogs (*volts*) available to move the sled (*watts*). He thought if he somehow could move it to the edge of the frozen lake, the *resistance* on the ice would be less and the dogs could probably move the sled once it was in motion.

His neighbor had two donkeys (*volts*). Each of them could easily move 600 pounds (*amps*) to gain the desired result of moving the

In Series
60 Volt X 20 Amps

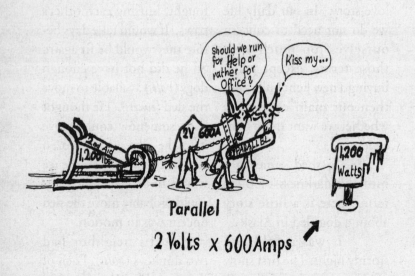

Parallel
2 Volts x 600 Amps

1,200 pound sled ($2\textbf{V}$ x $600\textbf{A}$ = $1,200\textbf{W}$). This solution, however, would not come without its price. In addition to the fact that he would have to pay his neighbor to rent his donkeys, the path to the lake was very narrow and the two donkeys could not possibly maneuver it safely walking next to each other. They would get stuck between the big boulders, and there was no way to arrange them behind each other. A wider path was needed, or. . . . There was another solution.

He was breeding sled dogs and was proud of having the biggest selection of them. In all he had 60 young dogs. Each of them could probably manage to pull 20 pounds. Would that be enough? He calculated: 1,200 divided by 20 = 60. ($1,200\textbf{W}$ ÷ $20\textbf{A}$ = $60\textbf{V}$) Yes, that would be enough—it was worth giving it a try. Plus, they easily could manage the narrow path.

Any teacher of electrical engineering reading this story would probably hang himself in frustration, because electricity is not that simple; it is certainly not dogs and donkeys. Plus the animal protection agency would probably find a few violations, too.

But, what we can learn from Tom's night in the wild is that we can exchange two variables, *volts* and *amps*, to gain the desired result, or, if we cannot exchange them, we will have a lesser result in *watts*.

The electrical engineering professor who hanged himself will most likely mention in his farewell letter that one of his reasons was my total mishandling of the term "resistance."

In the first example, the lesser resistance on the ice referred to a *physical resistance*, a resistance you encounter when you want to start your table saw or your blender filled with frozen bananas to give your diet-drink a better taste. To start up this machine, you momentarily need a lot of amps (or, if not available, some strong donkeys) to get things going. Once the initial starting resistance is overcome, you need much less power to keep things moving.

In the other example, torturing those poor donkeys by squeezing them through the narrow path (I promise, Tom did not do it), was referring to a resistance that we encounter when our amps go up (two donkeys each pulling 600 pounds) but our cable is too thin (path is too narrow). When those donkeys rub their bellies against the boulders, they will get pretty hot and they may burn their skin. The same is true with cables that are too small. They will get hot and hotter and may even burn out.

Another grievance of our electrical engineering professor would be my loose interpretation of the term *volts*, which is actually the pressure applied and not the pull of two donkeys. But I could not possibly have these poor animals push the sled, when they are trained to pull it.

It is also not true that each voltage item has a des-

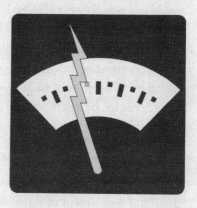

ignated amount of amps to it, like the donkeys. The term *ampere* (sounds French to me) refers rather to a flow of electricity, more like water through a pipe. We know that the load on the sled had to reach its destination in the shortest amount of time possible. Moving that 1,200 pounds in a certain time-frame would be the best explanation, or—using the above-mentioned water-in-a-pipe example—having a flow of three gallons of water per minute could give us a picture of three ampere.

Another term that we will have to understand, that derives from the term ampere, is "ampere-hours." All batteries are rated in *amp hours*. And while the term ampere can be compared with a load moved or an amount of water flowing, it is by real definition referred to as "per second." And now, hold on to something—here is the official amount of electrons that are moving per second when one ampere is measured: 6 million million million.

Now, if we use one ampere for one hour, we use one *amp-hr*. If we have a battery that is labeled for 220 amp-hrs, we can either use one amp for 220 hours or 220 amp for one hour, and of course all the possible combinations thereof.

In this chapter we learned what electricity is.

(When a man loves a woman. . .or when a priest

talks to an electrician.) We know that 12 dogs (*volts*) can be substituted for 2 donkeys (*volts*) in order to move the 1,200 pound sled (*watts*), if those dogs can each pull 100 pounds and if they don't bite each other. We learned that a narrow pass (cable) can restrict the two donkeys from passing through but that they can be replaced by 60 puppies. And how much must each puppy pull? 1,200 ÷ 60 = 20. We also learned that time is a crucial factor if you only have a limited amount of power available. If you have a full battery of 220 amp-hrs and you want to use 10 amps, it takes you how long to empty the battery? 220 ÷ 10 = 22 hrs. Now here is the $10,000 question. Using the same battery of 12 volts and 220 amp-hrs and burning a 24 watt bulb, how long can you burn it? Remember, we need to know the amps of the light bulb. The watts are the resulting outcome, like the weight of the sled. (**Watts = Volts x Amps**) If we want to know the amps, we divide watts by volts. (24 ÷ 12 = 2) Two amps can burn for 110 hours to empty the battery.

Congratulations, you are now entitled to enter the drawing for the sweepstakes. But after you have read the fine print, you may not want to be the winner because you get paid $10 annually for the next. . .how many years?

Chapter Four

HOW IS ELECTRICITY CREATED?

Bird On The Wire

Have you ever wondered why such a fragile thing as a bird can sit on a power line that probably holds around 300,000 volts? Well, I have, and it led to the question, "Why don't they get toasted?" And where does the power come from, and which direction does it flow in, and how did it all begin? Can electricity be created?

Even though we know that power companies do it, the experts want to make us believe that nobody can create electricity. They claim that the universe has a limited amount of electricity which cannot be destroyed or created. And it all began with the "Big Bang." (Remember this the next time you plug in your hair dryer.) All we humans can do is manipulate the existing electricity by shifting it—moving it around to create electrical phenomena. I was so glad to hear that because it puts the power companies in a different perspective.

Disregarding reports in the Old Testament, I want to disclose a few more recent historical events that led to the exploration of electrical

phenomena. How about frog's legs? Some people swear by them—served with a white wine sauce. To scientists, however, they were the entry into the age of modern electrical power. It was not so much the taste, but rather their continuous movement after they had been cut off the poor frog, which led the scientist (I forget his name) to wonder why. And to make a long story less dramatic, electricity was found to be the culprit.

The next thing we know is that Thomas Alva Edison lit up his first light bulb and, about a day or two later, built his first electrical power plant in New York. But if you remember your high school physics class at all (besides the fact that you discovered the laws of aerodynamics when you finally managed to calculate the right arc to throw a slip of paper into the lap of your future sweetheart which led to your dream date where you explored electrical phenomena in person), you may remember that it is magnetism that really brings things together.

Magnets are always mentioned in connection

with electricity. The exciting thing about magnets is that they are one of the best analogies for a society. Let's say we have the most radical conservatives on one side and the most liberal socialists on the other. In the middle, you have the undecideds or the "lukewarms." Now, one side has money and power and the other side has not. You can imagine that there is quite a powerful tension between them.

Both parties locked

into one room create such magnetic fields that anybody moving through this room would be electrified.

Power companies know that and use this phenomenon not just to lobby in Washington but also to make money. They build big machines in which big coils of copper wire move through strong fields created by big magnets. They call these machines *generators*, even though they should know that electricity cannot be generated (or created) but only shifted. But, at times, they feel a little bit god-like after all because, at least until now, they have had the monopoly in their area.

The difference between today's power plant and Edison's can be expressed with the terms, *AC* and *DC*: *Alternating Current* and *Direct Current*. If you remember our room full of magnetism, the most extreme parties will gather on each side of the room. Let's call the ones who have it all the plus side and the ones who don't the minus side. If a fairly neutral person, let's call him Colonel Copper, moves through the room from plus to minus, he will first be agitated by the strong opinion fields of the ones who have and will start to believe that he also has. Now, moving to the other side and being agitated by the minus fields, he will lose what he just gained. One can say (besides calling him an opportunist) that something moved in him from plus to minus.

Well, Edison moved a copper rod through the fields of a magnet and electrons—so scientists claim— moved from one side of the rod to the other. It moved from plus to minus and electricity was born.

Now, remember the poor electrical engineer in Chapter Three? He would argue that electrons actually moved from minus to plus, because they are by nature negative and hence the surplus must have been on the negative side. But, just to make things easier for the common people to understand, where a surplus is called "plus," he would agree that electricity flows from plus to minus. (Or did they just call the minus plus and the plus minus?) However, the flow from one side to the other is called direct current or DC.

Since it is much easier to turn things than to move them back and forth, some smart person took a coil of copper and turned it in a cylinder whose walls were lined with magnets. Later they put the coils of copper on the outside and the magnet on the inside. The effect is the same: whenever copper moves through magnetic fields, electricity starts to flow. The only side effect of choosing a round environment was that it created a constant flow of electricity, changing from plus to minus and back, without interruption, in a pulsating frequency. It was

30

going up and down and up and down so quickly that our Colonel Copper could not tell anymore whether he had or had not. He was alternating his opinion so fast that, from then on, he was referred to as Colonel *AC* or *alternating current*.

The amount of time in which the change from plus to minus happens is called the frequency or the cycles per second, which is measured in *Hertz*. (Named after Heinrich Rudolf Hertz, a German physicist who produced the first radio waves artificially.)

Disregarding all the possible interventions from our electrical engineer, let's say the following: AC is easier to generate, it can be easily transformed into high and very high tensions (and back down, of course), and hence can be more easily transported. Remember, two donkeys need a bigger path or cable than 60 puppies. That's why we send puppies through high tension lines.

We also know now that electricity flows from plus to minus (or sort of) or call it from *hot* or *positive* to *ground*. We are walking on the ground and the bird sits on the wire. Besides our admiration, nothing flows between us. If, standing on the ground, we would attempt to touch the bird, we would become a conductor between hot and ground and, unless our name was Colonel Copper, we would look like a forgotten piece of toast in the toaster oven.

Now, you might ask, if AC has so many advantages, why use DC at all? Remember the big blackout in New York City in the 1970s? Yes, the one that was followed by a small baby boom nine months later. The problem was that several power plants failed simultaneously and the rest shut off because they could not produce enough electricity to make up for the loss. Why did they not store enough electricity as a reserve, one might ask? Well, the only flaw about AC is that it cannot be stored. **Only DC can be stored.** That's why cars, airplanes, railroads, and photovoltaic systems use DC.

In this chapter, we learned about bird watchers, and why they can turn easily into a piece of toasted Wonderbread if they decide to touch the bird and become a conductor between hot and ground.

We also learned about politics and about

spineless colonels. If all politicians would change their opinions at 60 hertz per minute, they at least would be good for something.

We also found out why power companies think they are god and that it is time to break up these monopolies or, even better, be independent from them.

We also learned why they use AC, which is more flexible than DC but lacks one quality: it cannot be stored.

Chapter Five
HOW DOES SOLAR ELECTRICITY WORK?
The Sunburn

One day—I do not remember how old I was—I was at a beach. It was vacation time and the sun was shining. It was warm, the waves created a rolling sound, and I was lying in the sand. The next thing I remember was that I woke up, burning hot in my face, and red all over my body. When I looked in the mirror that night, I could not believe it. My appearance had totally changed. And I wondered as I suffered through the after-effects of that burn, what had happened? How can the sun have so much power, and what happened to my skin? I was not yet going to school and had no scientific training, but I understood that things react to sunlight.

After I grew a new skin, I became more careful and started to observe my surroundings in a more scientific manner. I observed that small puddles of water dried out in the sunlight, and, so it seemed, did old faces as well as my lunch bread while the butter on it melted and dripped on my new shoes. My naked legs hurt when they came in contact with a hot car seat, and the big boy from next door demonstrated the biggest miracle of all by holding a magnifying glass to dry straw and it caught fire!

Later, in school, I learned about many more wondrous things the sun could do. For example, I learned that sunlight hitting a green leaf can, by process of photosynthesis, make the leaf use carbon dioxide and produce oxygen, which we need for breathing. That's why we want to keep our trees.

As we learned earlier, a certain amount of energy can be channeled to create a certain result. Dogs properly trained can get us ahead. So can donkeys, even though they tend to be more stubborn. Our sun puts out an inconceivably high amount of energy—an estimated 100 trillion kilowatts. Some of it reaches the earth. At noon, a 10 square foot surface gets hit by 1,000 watts of sun energy. "If we only could we surely would . . ." put it to some use. Well, we can, but only a very few people do.

Remember those old light meters for cameras,

used as early as the 1930s? Well, they were the first photo cells. The principle has been known for almost 100 years and has not changed much since. If sunlight hits a photo cell that is layered with one coat of silicon that has a positive charge and one layer that has a negative charge, it acts like a mom turning the lights on at her teenaged kid's party. Things start to move in this room, and almost everybody would have to trade places. As a result of this, a small electrical current is created. Needless to say, after mom leaves the room and someone has turned off the lights again, the process will certainly be reversed. Now if we would connect several of these teenage party rooms (solar cells) in a row, we would create what is referred to as a *solar panel* and several panels create what is called a *solar array*.

The electrical effect of this principle was already used in the space program of the 1950s. The process of getting electricity this way is called *photovoltaic*. Since there are no revolving magnets involved in this process, we can rightly assume that the electrical current we get is DC. The photovoltaic process converts light into DC. While a sunburn is a pretty hot process, every teenager would agree that this photovoltaic process is rather cool. In fact, heat hinders the process. (Of course, someone may now argue that the real damage of a sunburn is not the result of the heat but of the UV-rays, but it only shows you can't make everybody happy.)

It's the light that makes these electrons move between the two layers of silicon which creates an electrical current and not the heat. In fact, you get a higher charge rate in cooler climate zones. Of course, as already indicated, this process can be reversed when there is no light. At night, some of the energy we received through our panels and stored in our batteries would, by reverse flow, dissipate into the universe. But there are electrical or electronic "taps" available that shut off the flow at night.

The question may arise, if it is that simple to move electricity, then why isn't everybody doing it this way? What about the power companies, and why are they using polluting resources, like oil and coal, when this

clean option is available? To answer these questions, we would have to get into politics. I'd rather put two donkeys in front of the sled than argue with a politician. It can be done, has been done, and will be done. But don't underestimate the power of lobbyists. And remember the Gulf War? The reason why we are using polluting resources is as polluted as the resources.

Just take a drive through the Four Corners area in the Southwest. Around Shiprock, the air gets so dense you wonder why

you left Los Angeles in the

first place. And just because somebody burns the dirtiest coal possible in one of the sunniest parts of the country to get electricity.

In this chapter, we learned that even a sunburn can have good aftereffects and that the teenaged love life is a cool thing because it gets the energy flowing. But the same process can be reversed if there is no light.

We also learned that the solar resource was already fully developed in the late 1950s but that a big pile of dirt would have to be moved to make it available as a real resource in today's polluted times.

But we also know now that we do not have to wait for this to happen. We can clean up our own backyard any time. We can install our own photovoltaic system. Whether this is a lucrative option if you live in suburbia is another question.

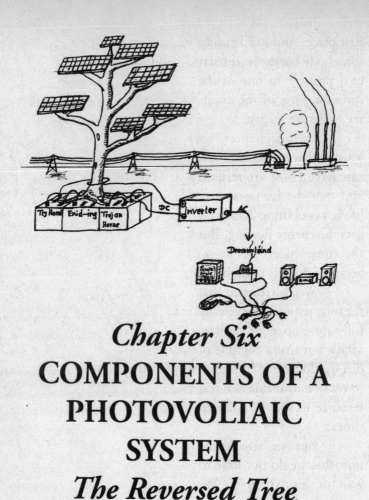

Chapter Six
COMPONENTS OF A PHOTOVOLTAIC SYSTEM
The Reversed Tree

If we install our own photovoltaic system, we become our own power company. One may think, "Now, that's great. Let's create more energy than one needs and sell the rest back to the power company." And, in fact, that is possible. During the more clean-energy-conscious years, a law was created that required the power companies to buy back electricity. Unfortu-

38

nately, they did so at a wholesale rate but sold it at a retail price. In effect, you got much less money selling your electricity than you paid when buying it. Not a profitable option unless you enter the game big time. However, recent changes require power companies to reimburse you at the same rate at which you buy electricity, by simply letting your meter run backwards (called *net metering*). Only after reaching the zero-point will they reimburse you at the wholesale rate. This, of course, may be subject to change at any time. (If you are serious about it, ask your local authorities before engaging in such a venture.)

But even owning your own little power company and being self-sufficient can give you a great feeling of accomplishment. With it comes a new awareness and, of course, new responsibilities. You consume what you "create." Your resource may be limited by your budget,

so you will need to stop wasting energy. To some that may seem like a sacrifice, to others this may lead to a simpler and more efficient lifestyle.

Knowing that the USA uses 40% of the world's resources, it may be time to wake up from your beauty sleep with three-car families, microwaves, curling irons, air conditioners, and electric can openers. You will need to go through the process of asking yourself: What do I really want and what do I really need? And this process may be an ongoing one. But it certainly needs to reach some sort of resolution when you start planning your system. Every system from the smallest to the biggest has the same main ingredients, which are what we want to talk about in this chapter. If you plan a small system, you may skip some of the fancy parts, but you still have to cover the basics.

Since electricity is such a theoretical subject

and most of its applications are hidden behind walls, let's look at some more organic analogies. Let's look at a reversed tree. Yes, a tree that collects energy from the outside and puts it into the soil so that other plants can grow. Farmers know that certain plants can collect nitrogen from the air and put it into the soil. If their soil is poor in nitrogen, those farmers plant these plants between seasons or with the crop to provide the nitrogen that other plants need to grow.

Let's first look at what parts of the tree are needed to do the job and then, in the next chapter, look into what size tree we need to feed all the plants we want.

A. THE ARRAY (The Foliage)

As a first definition of the term "array," we find in *Webster's Encyclopedic Dictionary*: "To arrange or draw up in battle order." Well, it seems that the military has dominated much more than just dealings in real estate. But all is not lost. We find as the last definition: "a regular arrangement of antennas. . ." It seems that the definitions have been arrayed in historical order, in all a pretty systematic line-up, just like troops.

All the word implies is that we line up something or someone either in order or by order, and we have an array.

Our array consists of a number of solar panels, which as we know also consist of a number of small cells wired together. We also know by now what these cells consist of: teenagers having a party in a dark room.

Panels come in all kinds of sizes, voltages, and watts. There are small ones used for recreational pur-

bucks per watt. As far as the durability and performance are concerned, I can say that I have not yet seen or heard of a solar panel going bad. (Except if you take a sledgehammer to it.)

But there are differences in construction. One type is called the *single crystal cell construction*, which the majority of panels are made of. Small single silicon crystals are grown and those cells are connected together to form one solar panel. A different system is the *amorphous film technology*. These are flexible, lightweight solar cells deposited on a stainless steel substrate, which means the whole panel is one cell. The efficiency of these panels is less than single crystal cell modules but they may be the future because of their flexibility. Just imagine that your roof shingles are all little flexible solar cells. This would make your whole roof one big array. But until they become more efficient and

poses to beef up car or marine batteries which are designed for 12 volt systems. There are big and powerful ones for people who like big and powerful systems, and who have big and powerful resources. And there are recycled ones that have been taken out of solar power plants, refurbished, and thrown on the market at more affordable prices.

Whichever one you choose depends on budget and space considerations. The older recycled models tend to be less efficient, which means they have to be bigger or you have to put up more to get the required watts. You need more space per watt but you pay less cheaper, the single crystal cells will decorate your roof or backyard.

Almost all systems consist of more than one of these panels connected with each other either *in parallel* or *in series*. Now, you hate to hear another electrical term, don't you? (I try to spread them out a little bit so you don't lose it right at the beginning.) If you remember the story about Tom and his dog sled, we had our dogs lined up behind each other and one seemed to pull the other. We see a *series* of dogs who perform the task. If each of them is capable of delivering one volt, lining them up in a series of 12 will produce 12 volts.

Now, the two donkeys, big and fat and next to each other, pull this sled *in parallel*. Each one can deliver 12 volts and, if they pull as a parallel force, they pull as a 12 volt pair. If we decide, for instance, to buy some recycled panels that come in 4 volts each, we will have to connect four of those in series to charge our 12 volt batteries.

Why, you will say? Four times 4 is not 12, but 16. You got a point there, and I am glad you are staying on top of things, but here is why: Your beautiful aquarium needs to be emptied, but it is much too heavy to lift or tilt to pour the water out. You have no pump available either. What you can do is siphon the water out of the container. You do this by placing another container at a lower level than the aquarium. Then you use a hose and stick one end into the aquarium. Next, you lower the other end of the hose below the water level of the aquarium and suck on it until the water starts to flow. Once the water starts flowing, it will continue to flow until the higher container is empty or until both water levels are equal. In other words, you need a higher water level to fill the lower container.

This picturesque example applied to our batteries means that you need a higher voltage than your indicated battery voltage to fill up your batteries. A 12 volt battery, for example, is considered full at a level of at least 12.8 volts. So you need a higher voltage source to fill your battery bank. Car batteries get charged with 14.6 volts; 12 volt arrays, with between 16 to 20 volts.

So, in order to reach this voltage, you need more than one panel of 4 volts each and you need to do what? Right, you need to connect them in series. Here are the electrical terms for series and parallel connections: If you connect in series you go from the plus of one panel to the minus of the next panel and from the plus of this panel to the minus of the next panel until you reach the desired voltage.

If you connect in parallel, you go from the plus of one panel to the plus of the next panel to the plus of the next panel and so on. You do the same with the minus. Between all the pluses and all the minuses, you still will have the same voltage that you started out with.

If we had one panel producing 16 volts, why would we want to connect more than one panel? For the same reason that you would want to use two donkeys in front of the sled—because one might not be strong enough to perform, to deliver the required watts. One panel may not have enough amps available. But we will talk about sizing your system later. (Did you notice that I am good at postponing? Must be the southern sun of New Mexico.)

These solar panels, arrayed together, are our collecting "foliage." They absorb the energy the sun makes available. Connected in series or parallel to each other, we can collect enough power to refill our batteries if we sized our array in proportion to our battery bank.

We learned that our tree foliage needs to be the right size in order to send enough energy down into the soil (to refill the batteries after use). That's why we may need to connect several panels either in series or in parallel or in a combination thereof.

donkeys pulling next to each other (they can't bite their tails).

In case you choose a 24 volt system, you would choose a combination of series and parallel. And, to make this example more interesting, we will use mules or asses. You pair them up

In Series
60 Volt × 20 Amps

Parallel
2 Volts × 600 Amps

In series means several dogs pulling in line, each mouth (*plus*) biting the tail (*minus*) of the one in front.

In parallel means two

(*parallel*) and put two pairs behind each other (*series*), so they can bite each other's tails.

B. THE CABLES (Twigs and Branches)

You might think there is not much to be said about cables in a solar system. All they do is get tangled up with each other, looking ugly and always being in the way of the important components. It seems at times that it would be best to do away with them altogether. And I wish one could, because too many things depend on those cables and too often they are the cause for malfunction, poor performance, or sparking and fire. Every aspect of the cables is important: the size, the material they are made of, the length, and, last but not least, the color.

Let's first look into the siz-ing of wires. If you open the hood of your car, you might notice that there are cables all over the place. But two things might strike your eyes—the variety of colors and their size. Almost all of them are very thin. Now, you might think, of course it's only a 12 volt system, so you can use thin cables.

Those were the exact words of a customer who called for help late one summer afternoon. He had added some solar panels to his existing ones. He had made careful calculations: 4 more panels, each delivering 25 watts, should give him an additional 100 watts. He also knew the formula to

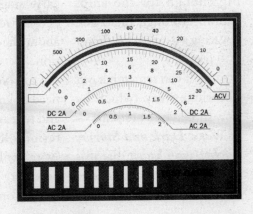

translate this into amps which then would be reflected as additional gain on his ampmeter. He came up with an additional 6 to 7 amps. But when he looked at his ampmeter, he only noticed an increase of 3 to 4 amps at peak times.

My first question to him was: "What size are your wires coming in from the solar panels?" And there it was, the stare with a big question mark behind it. "Look pretty thick to me," he said, "a big flat and gray wire."

Once I had a look at them, it turned out that the wire was a number 10-2 UF which turned out not to be big enough for the amps running through them and covering a distance of almost 100 feet.

If you remember our notorious story (which by now should be a bestseller— *It Was a Dark and Stormy Night*), you know that it is of tremendous importance that you use the right size wire to let all the electricity move through it without rubbing itself and turning your wire into a space heater. But not only the size of the wire is important but also the quality of its insulation.

In many cases when I was called to service or upgrade an older system, I noticed that the wire used was not only inadequate as far as its *size* was concerned but also as far as its *type* was concerned. It looked like an old dried-out lake bed. The insulation was weathered and partly broken off so that, at times, the conductors started to touch each other and shorted out.

It seemed that, in the early stage of solar power, people were rather ignorant about the details of installation.

Now what do we need to know about the size and types of wires? First, the terminology: Of course, "wire" only exists along the periphery of your property, in the form of barbed wire

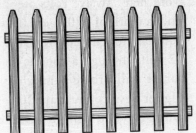

or fence wire. Even though there is such a thing as a wiring diagram which you can use to wire your house or your solar system, you still need to use either *cables* or *conductors* to do the job.

There is cable containing several conductors which, according to their colors, have different names—and sometimes different sizes. You can size your wire by using a formula which incorporates all the givens, like amps, voltage, distance, resistance of the conductor, and the percentage of the volts you lose running electricity through them. It is easier, however, to use tables which have been created to give you the right size of conductor for your specific voltage and distance.

(Although I promised not to include any graphs and charts, I will include a "table" for conductor sizing. See **Appendix D,** p. 204.)

The conductor size is measured in numbers of *AWG (American Wire Gauge)*. (Strange isn't it, there is the word "wire" again.) Type and size have to be printed or engraved on conductors or cables. The smaller the number, the bigger the conductor. Telephone wire comes mostly in size 18 AWG. Your house wiring is mostly size 14 or 12 AWG. You electric clothes dryer uses number 10 AWG and your electric range number 6 AWG. Electricians use a rule of thumb: number 14 is for 15 amps, number 12 for 20 amps, number 10 is for 30 amps, number 8 for 40 amps and number 6 for 50 amps.

Since many people are aware of this rule, they try to apply it also to their solar equipment and that's where they err. Remember (I know it's hard, but please try) the equation we used earlier—the one and only? Yes: **Amps x Volts = Watts**. Now you also remember that, if you change one value of the equation, all the others change as well. A similar formula exists with regard to the resistance of the conductor. I should not give it to you because this course is only Photovoltaic 101, but since you stayed with me until now, here it is: **Resistance = Voltage ÷ Amps. (R = E ÷ I)**

Now the same is true for every formula. If you change one value, all the others change as well. In the solar DC application, you may remember we use a voltage that we usually do not use in a conventional house, which is either 12 volts or 24 volts. If you consider the above rule of thumb that electricians use, you will immediately see that what works for a house voltage, usually 120 volts, does not work here. Twelve volts are only a tenth of 120 volts. If you want to send 20 amps at 12 volts through a conductor, it will create a much higher resistance than 20 amps at 120 volts. Of course, at a low voltage, the distance becomes a crucial factor. Every foot counts because you have so little pressure. Now you also know why high tension lines can go long distances. The higher the voltage, the longer the run can be without too many losses.

In the case of the customer I mentioned earlier, it meant he needed to upgrade his conductors to #6 to cover the nearly 100 foot distance.

Now that we have explored the sizing of the wire in depth, let's talk about *temperature* and the appropriate *insulation*. What needs to be said about temperature? Well, some like it hot. But most don't. Conductors are also rated for maximum temperatures. These are temperatures created from the inside of the conductor as well as by environmental conditions. You know quite well that, on a hot summer

day, you can fry eggs on a tin roof. (That's how the solar cooker got invented.) If you throw a piece of electrical wire into your solar cooker, it may add a nice variation to the taste of your soup. But, imagine the cables you threw carelessly on your roof when you "temporarily" connected your solar array, frying in the hot sun for several seasons. By now, they might have lost all their insulation and are starting to divert the electricity coming from your panels to your tin roof.

What does the temperature of a conductor mean and where do we find the information on it? Well, it is all written on the conductor or cable. Remember the customer who called me mentioned a 10-2 UF cable. If you look at the printing on the cable, it might look like this: *Size AWG 10-2 UF-B Sunlight Resistant.*

The first number (*10*) refers to the *size of the conductors.* The second number (*-2*) means that there are two *conductors* in the cable. The letters *UF* stand for *underground feeder,* which means the cable can be buried directly in the ground. The next letter (*-B*) refers to the *temperature* the conductors in the cable are rated for, which in this case is 90°C (degrees Celsius). And the words following mean that the cable is *sunlight or UV resistant.*

Very well, you say. So I can just throw them on my tin roof—it can't be that hot up there. You may be surprised. I have never measured the temperature of a tin roof on a hot summer

day, and 90° C is about 200° Farenheit, but I am sure it will come pretty close to this temperature. And here is another consideration. The -B is only indicating the temperature the conductors can take. It doesn't indicate what the outer insulation of the cable is rated for. For example, on a different kind of cable, well-known to many as Romex cable, it may say *AWG 10-2 NM-B 60C.* (Rumor has it that the name originated with cable coming from a place called Rome and shipped by Fedex.) This wire, although of the same size as the UF-B, is not rated for outdoor use at all. The letters *NM* indicate that it can *only be installed inside* framed walls and ceilings. Even though there is a -*B* behind the letters which indicates that the conductors can take heat up to 90° C, the following number (*60C*) indicates that the outside insulation is only rated for 60° Celsius.

Now let's talk about the *type* of cable or conductor. As we know now, *UF* means *underground feeder*. It means it is resistant to soil, moisture and—hopefully—to the teeth of the creatures in the underworld. If you want to make sure that this does not get put to

the test, run your cable or conductors in PVC pipe. This is especially important when you have to use a conductor size bigger than #6, which is the biggest size UF cable that is manufactured. Everything bigger than #6 comes only in single conductors, which means one single piece of copper (solid) or several twisted together (stranded) and surrounded by one layer of insulation. Here you have the choice of *UF* type conductors or *USE* type conductors for direct burial. Again, when it comes to single conductors, it is advisable to run them through PVC pipe, underground as well as above ground.

A few words on the size of conductors: As mentioned earlier, the smaller the number, the bigger the wire. Usually they come in even numbers. Only when you get below #4 do they count down to #1, using both even and odd numbers. Past #1 it continues with 1 O/D (say "odd"), 2 O/D etc. A pretty common conductor size in a solar system is a 4 O/D conductor, running between the battery bank and the inverter. This conductor is bigger than an average thumb. If you remember our calculations in the beginning, you may know why you find a conductor of this size. If you have a 4,000 watt inverter and a 24 volt system, you can calculate that it takes 160 amps to deliver the 4,000 watts. (4,000 ÷ 25 = 160). Now consider that the inverter may be capable of a surge of up to 10,000 watts and that you need 400 amps running through this conductor—it therefore requires a 4 O/D conductor.

Now you know much more than most of the solar installers do, as far as cables are concerned. I cover them in such detail because, in 90% of all the installations I have seen, the wrong size or type of cables has been used.

At least now you know that the writing on the cables is not all that difficult to understand. You also know that expensive equipment can be rendered useless by using twigs and branches that are too thin.

C. THE CHARGE CONTROLLER (The Tree Trunk)

But back to our more organic environment. Whatever energy the leaves collect needs to be channeled into the ground (the batteries). The branches of the tree represent all the cabling, a very important part of every system. The trunk, however, gives direction and regulates and controls the flow of energy.

While soil may be able to absorb an almost infinite amount of energy, our battery bank cannot. When our 12 volt batteries reach their fully charged level at 12.8 to 13 volts, that's all the food they want or need. Someone has to turn the tap off. This someone is called the *charge controller*.

Modern charge controllers do, of course, much more than that. Batteries are not easily satisfied. If they reach their 12.8 volts fully charged stage, you may think they are fat and happy. But if you check a little while later, you will find that their voltage has dropped without anything having been turned on and using energy. So you start charging again until they are full, and the same thing will happen over and over again. This is the job of the charge controller: charging, checking, and charging. The charge controller actually is smart enough to give the batteries more than they need. It overcharges the batteries usually up to 14.4 volts, then turns itself off and checks again.

This bad behavior of the batteries is called *float charge stage*. It is called that because only the surface of the batteries, the layer floating on top, got charged and quickly reached a high voltage. But it did not "sink in." The deeper layers of the batteries remained at lower voltages. Repeating the charging process over and over will slowly fill the batteries up to their capacity. This process is called *trickle charge*. It's like pushing your sleeping bag into its pack sack. At first you make good progress, but the last bit you have to push and push until finally the pack sack is totally filled with your sleeping bag. Now you better close it fast before it pops out again. Preventing this from happening is another function of our charge controller.

If you remember, plants, by process of photosynthesis, transform carbon dioxide into oxygen as long as sunlight is available. At night the reverse will happen—they will use oxygen and release carbon dioxide. A similar process occurs within our solar array at night. It releases or discharge electricity into the universe. That explains why in a dark room full of teenagers, there

is a lot more action going on before mom turns the lights on.

Our charge controller makes matters simple at night; it turns itself off and prevents any back flow of

electricity. In the morning at the first ray of sun, it awakens again and the process starts all over.

If you have used a lot of energy during the night, the charger starts in the morning with a good breakfast for your batteries. It shoves the energy into them as if it were bulk food. That's why this stage is often referred to as the *bulk charging stage*. This is the principle of a German diet: a hefty breakfast (lots of sausage), the main course for lunch (usually eaten hot), and very little for the night.

Sophisticated controllers give you digital readouts of their continuous efforts which may include the status of your charge in amps, the voltage of your batteries, and possibly indications of usage.

That is about all I can say about this little black box. If it works, you do not think about it much, but if it fails, you may be sitting in the dark because your batteries won't get charged. Or there is the possibility that the controller will fail in the "ON" position, which means it keeps overcharging the batteries, leading to battery damage. (More about this in *Chapter Twelve, Troubleshooting*, p. 171.)

Even though it is small in appearance, the charge controller has an important function in your solar system. Never underesti-

mate the power of little dark boxes.

It is also quite clear that, following the principles of the German diet, with the bulk during the day and just a trickle later on, will top off your batteries just fine.

D. THE BAT-TERY BANK (The Soil)

Now we come to the most misunderstood part of a solar system: the batteries. Again the word can imply many things, such as an array of guns used together on a war ship. It also has a few other mean meanings. But let's look on the bright side of life, to the light and other useful things a battery can give us.

We all know those little round things that lit up the first flashlight we got for summer camp. Or the even smaller ones that keep our digital pagers from disconnecting us from the world. But the really big and heavy ones make it possible for us to get our favorite motor vehicle to start, even on the coldest winter mornings.

These are two different types of batteries: the former one is called a *dry battery*, which is sealed and usually gets thrown away when its power has been used up (they can be recycled, of course), and the latter one is called the *wet battery*, which usually is not sealed and which is designed to be recharged. (I know, I know, there are little rechargeable ones and there are big sealed ones—I'm just trying to simplify things a bit.) And there is a third type of battery, the *gel battery*.

In solar systems, you normally find a wet battery. These batteries are also referred to as *lead-acid batteries*. They have been around since the beginning of time and have improved very little. Basically, several negatively and positively charged lead plates are immersed in sulfuric acid. And the same principle that runs the whole known universe comes to work here, too. One side reaches out for the other, plus wants to reach minus. The acid is the mediator and helps communication between the two sides. This flow of energy between the plates is called electricity and, as we know by now, it is DC. A battery usually consists of several cells, each cell having a capacity of 2 volts. These cells, connected in series, form one battery with a total of either 6 or 12 volts.

This 100-year-old concept of a battery has the advantage that, once all the differences between the two sides have been equalized, and there is nothing left so say (i.e., the battery is discharged), the process can be reversed by putting new energy into the battery and, equipped with new arguments, it can start all over again. This is called a *re-chargeable battery*.

As I hinted earlier, there are various types of lead-acid batteries. The type we are interested in is called the *deep cycle battery*. This term does not refer to aspects of the female anatomy nor does it have anything to do with the collective unconscious. The term rather refers to the battery's ability to be fully discharged and recharged up to 500 times. The speed at which you discharge the battery will determine how long the battery lasts and how much power you will get out of it.

Depending on the battery's quality and the user's need for electricity, the battery can last from five to eight years. (Although I have seen battery banks as old as ten years.) Considering their weight, this is

a rather consoling fact. (After you unload your first set of batteries, you will know what I mean.)

Deep cycle batteries come in all sizes. They can be reasonably small and compact for marine applications or they can be gigantic units with each cell weighing several hundred pounds.

The most important thing to us is their label. Somewhere on it, you will find how many *amp-hours* the battery has. One amp-hour, as we recall, means that a light bulb using 1 amp can burn for one hour. A very common deep cycle battery

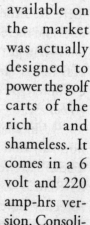

available on the market was actually designed to power the golf carts of the rich and shameless. It comes in a 6 volt and 220 amp-hrs version. Consolidating all that we have learned so far: if I have a light bulb that uses 6 volts and 1 amp, how long will it burn? (You will find the answer on page 666.) All I will say is that, after 9.16666 days, darkness will fall upon the reader.

Now, if we change the values of the equation and say that we have a heater that uses 220 amps, how long would the battery last? A simple calculation will reveal that 220 amps used for one hour equals 220 amp-hours. So our battery would last one hour, right? Wrong!

It all would have worked out nicely if it were not for this scientist named Peukert who in 1879 determined that a rapidly discharged battery will lose a lot of its power. In our case, it would lose up to 50% of its power. Ratings for deep cycle batteries are based on a 20 hour discharge rate. In our case, we would only gain 110 amp-hrs. To make use of the full 220 amp-hrs stored in this battery, we can only use 11 amps at a time, which would last us 20 hours.

Let's get a little trickier. If you have a 12 volt light bulb which uses 1 amp, how many 6 volt batteries do you need to power up this bulb and how long will it burn? Let's talk dogs and asses. (I don't mean politics.) In order to get a 12 volt battery, we will have to put two 6 volt batteries in series, plus to minus. The remaining plus and minus gives us 12 volts. The newly acquired 12 volt battery now has how many amp-hrs?

It still has 220 amp-hrs. Here is why: Each battery has 6 volts at 220 amps for one hour, which gives us how many watts? 6 x 220 = 1,320 watts or watt-hrs. If we double the capacity, we should get twice the result of one battery, right? That's true: $12 \times 220 = 2,640$ watts. Our amps have not changed but the volts have. The result is twice the watts of one battery but the amps remain the same.

The light bulb we are using operates under the condition that it needs 12 volts to light up and it uses 1 amp. It will also burn for 9.1666 days or 220 hours. Remember we have not looked into what result we are getting, which means how bright both light bulbs burn. The first one, using 6 volts and 1 amp = 6 watts. The second one, using 12 volts and 1 amp = 12 watts. The second one will burn brighter.

Our above calculations are very theoretical of course. In real life, a deep cycle battery should only be discharged to 20% of its capacity which means you have only 80% of its charge available. This is an important

consideration when it comes to sizing the battery bank. In fact, you need to buy 20% more batteries to meet your needs in amp-hrs.

Now that we are so

$$E=MC^2$$

deeply involved in physics, it's time to talk about metaphysics for a bit. What is life and how does it relate to electricity? The answer is simple: without electrical phenomena there would be no life as we know it. All matter is based on the movement of electrons which are in fact electrical phenomena. (Hey, that was physics again. I am so sorry.)

Now that we know so much more about what life is, let's look again at our battery bank. (Yuk! Someone should clean the contacts, they are totally corroded. But we will get to that later.)

A battery bank or an

array of batteries consists of several batteries, or cells, connected either in series or in parallel to get the required amp-hours and voltage. The size of our battery bank is determined by our use of energy and by our charging capacity. If we had a big battery bank, we certainly could run a lot of electrical equipment, but once we have used the available resource, we will have to recharge it. Now, if we only had very few panels, we would constantly starve our batteries because we could never recharge what we used. That's why the battery bank and solar array have to be well balanced.

What would happen if we constantly starve our batteries? Well, batteries have a memory. And this is not a joke. If you constantly undercharge your batteries, you create an internal memory which is a layer of oxidized lead on the plates inside the battery. As a result, the battery will not charge to its full capacity any more. If we would let this process continue, the oxidized layer would slowly transform into crystals. Unless you plan to trade in precious stones, you want to stop this process.

Even a good-sized battery bank will eventually create a memory, because the sun isn't always shining and

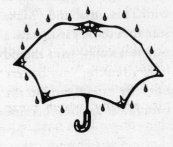

on certain days there is just too much on TV. Monday night football and three soaps plus your food processor running wild—let's face it, a thus-tortured battery will not forget. But there is a way to make it forget! It is called: *Equalization* or "cooking" the batteries. Don't get your big pots out yet. All we need is a sunny day or other means of charging the batteries. All modern charge controllers have a means of equalization which simply means bypassing them. Now we let the battery voltage go way beyond its normal cut-off voltage and keep it there for several hours. We will notice that our bank starts to "cook," which means it will develop a lot of gas which escapes with a hissing sound. Make sure that those gasses have a way to escape—you should have a vent for your battery box. Be careful around the batteries at this time because the es-caping gas is mostly hydrogen which, in connection with oxygen, is very explosive. So no checking the batteries with an open flame and, please, if you smoke, do it someplace else.

During the process of equalization, you will notice that the battery voltage rises pretty high, mostly above 15 volts. This "burns off" the oxidation on the lead plates inside the batteries and the batteries lose their memory. After a few hours of equalization, you can go back to normal operation. During this process, along with the gasses, we also lose a lot of water, which will have to be replaced. But more about this in a later chapter.

I want to throw out the question: Is it possible to live on solar without a battery bank? The answer is yes, and no. It is not possible to just run directly off the sun because, as we know, "the sun don't shine every day." But, if we had access to another power source, like grid power, we could choose an inverter that is capable of switching back and forth between the two power sources as needed, even feeding surplus power back into the grid. This certainly is an option to be considered, except for one fact. Should there be

an ice storm passing through,

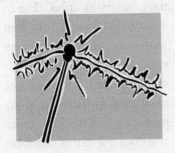

a tornado turning the power lines into shredded snakes, the inverter would be switching to candlelight power. For those cases, you would want to have a back-up generator, which in effect costs as much or more than a battery bank. So there is not that much of an advantage to living on a grid-inter-tied system without batteries.

We have learned that we can either be charged with battery or we can charge a battery. But we can only do this if we have a rechargeable battery. Our battery bank is our very soil which, if kept watered and energized, can give us a variety of electrical options. We also know now the true meaning of a deep cycle and that 500 of those are usually available in an average *deep cycle lead-acid battery*. We also learned that one does not have to play golf in order to own a golf cart battery, and that two of these make a 12 volt battery if connected in series. We also learned that, if we double our batteries, we get twice as many watt-hrs but keep the same amp-hrs, which in effect means we can burn a brighter light bulb with twice the wattage for the same amount of time or, if we are smart, we can burn a light bulb with half the watts for twice as long. Our optometrist will thank us for that, as we may have to see him more often.

But most importantly, we learned that a 100-year-old technology has survived almost unchanged into the atomic age. One wonders, is there a parallel to the combustion engine?

61

E. THE DC SYSTEM (New Sprouts)

As a reminder, D C means direct current and it flows from plus to minus, or, if you want to be correct, it flows from minus to plus (because the abundance of negative electrons which want to go some place else makes its location technically a negative pole). There are two characteristics of DC that can make our lives difficult:

1. With DC, you have to observe polarity carefully. If you feed reversed polarity into most DC gadgets (except lights), you simply blow them up.

2. The other problem is related to the wiring methods. If you remember that treacherous cold and stormy night when we discovered that two donkeys would not fit next to each other through that narrow path, and you apply that to your cable, you know that you may end up putting larger size cables into your walls. Edison's first power plant had to face the same problem. DC likes big wires and does not like to travel far.

And here, ladies and gentlemen is the technical side of that problem. The enemy we are up against is called *watts*, because it is the fixed item in our equation. Watts measures power, like horsepower. (Cars in some European countries are rated in watts rather than in horsepower.) A motor, a light bulb, or your DC radio needs a certain amount of electrical energy to give you the desired result. Each of our simple equations has three items or three seats available. One, *watts*, is already taken. The remaining two can vary. Or can they? Let's say for the sake of argument they can: A 24 watt light bulb that requires 12

volts needs 2 amps to power up: 12 volts x 2 amps = 24 watts.

But we could achieve the same with 6 volts and 4 amps, right? What we see is that, if the watts remain the same and the voltage drops, the amps increase. And an increase in amps means bigger cable. A too thin cable would heat up, because of the resistance the electrons would encounter passing through it, and heat can create a fire. It's that simple! So a lot of attention has to be paid when sizing the wire for your DC system.

Now you might ask, why then have a DC system at all, if we could have everything converted into AC? (Don't you hate rhetorical questions?) The problem with converting everything into AC is that it uses energy. The transformer of an AC system needs to be energized at all times, which in effect keeps your inverter on at all times.

There are electrical items that come with a cord and a little black cube at the end which you plug into your outlet, like answering machines, radios, lap top computers, etc. This indicates that these things most likely run on DC, and a lot of them run on 12 volt DC as shown on the labels on those cubes (which are called *transformers*). Using these machines as indicated would mean that we first transform 12 volt DC into 110 volt AC and then back to 12 volt DC. If that is not beating around the bush, I don't know what is. And every process of transformation requires energy which is usually lost as heat. In effect, we would go through all the expense to end up with something we started out with in the first place. Most of your DC equipment can effectively be run directly off your house voltage.

Water pumps are another good example of DC use. They are very efficient and, since they are usually in wet locations, they are very safe to use on DC. As a matter of fact, there are now very efficient DC well pumps available. Some pump only shallow wells, but a new model from Italy goes as deep as 600 feet. Powerful DC booster pumps can pressurize your water system. (See *Chapter Ten*, p. 147.)

Another DC application can be your refrigerator. S o m e D C fridges are very efficient. Others you can 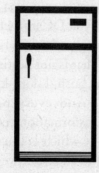 improve by insulating them well. Of course, if you plan to incorporate a DC fridge, you have to consider its use in your load calculation.

(You remember, the one that hits you over the head with a 2x4?) Although fridges have a low amperage, some of them run up to 50% of the day in the summer month. Hence, the better they are insulated, the less they have to run. (See p. 130.)

Another DC application is the smoke detector. Usually smoke detectors are interconnected with each other, so that they all go off if one detects smoke. They also have to be connected to the house power, which means they cannot any longer just run off individual batteries. The normal smoke detectors run off their own (dedicated) AC circuit. But this would mean that they would keep your inverter going at all times. Fortunately, there are now DC smoke detectors available that comply with the code requirements. (See p. 143.)

There are always some good uses for DC lighting, too. First of all, if your

AC inverter should ever fail (which rarely happens), it is good to have a few strategically placed DC lights around the house. (For example, where your broken

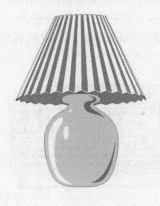

inverter is located.) A lot of modern designer lamps utilize halogen bulbs (see p. 150) which are usually run off a 12 volt transformer within the light. (That's why they are so heavy.) Having 12 volts available, you can bypass the transformer or buy them without a transformer and run them directly off your 12 volt source. (If you live in a 24 volt environment, guess what you can do with two 12 volt lights? Alright,

if you have two 12 volt lights and you connect them in series, they would work on 24 volts.)

Another DC application may be your source of heat. As mentioned earlier, the times of the lone wood burning stove as a heating source in a solar home may be over. Furnaces and boilers are now entering the modern solar home. Although fired with gas, they still need electricity to run all the controls, valves, and thermostats. If you ask your plumber, he will most likely say that all furnace and boiler systems run on AC. At least all he knows. Which means, he does not know it all.

There is a DC boiler system available which will avoid the trouble of having your inverter run overtime. Even if you do not find a DC boiler, there is another solution to this problem. But more of this in a later chapter. (Have you noticed that this is how I keep you reading?)

65

The Phantom (of the) Load

A *phantom load* is a load that is not real. It is a load that keeps your inverter on without a direct or useful need to be engaged.

Electronic equipment, like answering machines, cordless phones, TVs, stays on. (Hence the name *phantom load.*)

The electronic circuitry with its capacitors keeps sucking energy to their fullest capacity. That's because certain stand-by functions are built into these devices which need to be kept

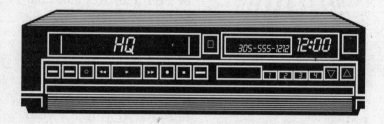

fax machines, VCRs, radios, CD players, computers, satellite receivers, clocking devices, electronic listening devices, etc., create, when left plugged into a receptacle, what is called a *phantom load.*

Yes, it is something that haunts you at night. It may not scare you, but you will definitely be able to hear your inverter buzz loudly and stay energized. Although you have turned all your equipment off, the inverter energized. (Your VCR, for example, will lose its memory when you unplug it.)

Because of the ability to create *phantom loads*, these devices need to be unplugged when not in use, or, if they can run off DC power, they need to be connected directly to your DC source. Many answering machines, cordless phones, radios, etc., come with a black cube, a device that you plug directly into your AC receptacle. This black cube is a transformer. Printed on it is technical data telling you its use in watts and its output in volts DC or AC. If the output voltage is compatible with your DC system, you can connect your device directly to a DC source. (*Note:* If you plan on doing this, be extra careful not to switch polarity accidentally because it will destroy your device. Needless to say, this voids the warranty. Also just cutting off the line between the little black transformer and the device and using it for your DC connection will get you into the same trouble if warranty repairs are necessary.)

Phantom loads need to be avoided because they keep the inverter running and waste energy unless you plan this energy loss into your load calculation. This

load can be considerable because inverters typically use between 6% to 10% of their requested output. A 4,000 watt inverter uses about 16 watts just to be on, which amounts to almost 400 watts per day and which, in effect, is the equivalent of an extra 75 watt solar panel. This can add considerably to the total cost of your system. (Today's price for a 75 watt panel is around $500.)

Radios, VCRs, or TVs, as well as computers

and other equipment that need to be run on 110 volts, can be controlled by running them through an inexpensive device called a *computer control panel*. This device has a bunch of switches up front. You plug all your equipment into its back and are then able to individually switch each piece of equipment "on" on its front panel as well as switch the whole panel off with a master switch. (In a later chapter, we will discuss some of these possibilities in more detail.)

Phantom load management is an important part of designing a system and can—at times—be a logistical nightmare.

We have learned that, even though smaller by nature than an AC system, the DC system gives us the first glimpse of our final electrical system. It can open the world of communication to us or at least give us good water pressure. We also learned that we do not have to leave in order to arrive, if our destination is where we already are. (I can't help it, but that sounds a lot like a political candidate hitting the campaign trail.) In plain English: why transform 12 volts to 110 volts and back again?

We also found out that water pumping, space heating, and beer cooling can be done with help from the DC system.

We also met a phantom. Not as charming and loving as the one from the opera, but rather a nuisance that we try to avoid. And we found out that we will have to read almost the whole book to find out more about how to get rid of this nuisance. (But if we got this far with the book, we might as well stick with it to the bitter end.)

At the very end of this chapter, there will be one final question to the reader: Which one is better, a 12 volt DC system or a 24 volt DC system? (Now this is a Zen-Coan, a riddle with lots and no answers. More about it later.)

F. THE INVERTER (The New Fruits)

I purposely deceived you when I used the words "to convert" for changing DC into AC. But since I had not yet educated you as to the meaning of inverting, you may have misunderstood the deeper meaning and inadvertently come up with the inverted meaning of what I intended to say.

Now, here is the definition of *Inverter*, fresh out of the **Webster's Dictionary**: "*Inverter*: one that inverts!" Time and again, I am impressed by the meaningful definitions in this book.

But, to be fair, I have to add that later the authors catch themselves by adding: "Device to convert DC into AC." Now, one might ask, if it converts, why not call it a "converter?"

Even I am shy when it comes to answering this legitimate question, except for the fact that a *converter* is a device that transforms a DC voltage to a higher or lower DC voltage. I can hear the question coming, if it transforms, isn't this a *transformer*? Of course not—a transformer transforms any AC voltage to a higher or lower AC voltage! I am glad that you learned something in this section before we even started it.

The *inverter* is the pulsating heart of our system. Everything we want or ever wanted can be run through or run off our inverter. It is a a device that takes our simple DC voltage and transforms it into a complicated and pulsating AC wave.

Want to know how it works? I won't tell you. All I will tell you is that inverters come in all kinds of sizes, forms, and applications. The simplest ones you can plug into your car cigarette lighter and then watch TV while you drive. The biggest ones are so smart that they can be programmed to cook your bacon and eggs in the morning, start your back-up generator, and take your kids to school. (I guess descriptions of the latter kind led the "professionals" of the solar industry to call my book shallow because they are vegetarians.)

Finding the right inverter for your system depends on your demands (you may be able to take your kids to school by yourself), and on the capacity of your battery bank, which as we know depends on the size of your array. (See how everything ties into each other?)

Almost all the newer inverters are smart enough to shut up when they are not needed, which means they do not drain your batteries just by being there, unless, of course, you run a phantom load. But as we learned in the last section, keeping the inverter running

with all its circuitry will add up to considerable consumption over a period of time.

Most inverters have an efficiency of more than 90%. This means they use up some energy for their service, fair enough. For this reason, you do not want it on all the time. So inverters are designed to have a *stand-by function*. The moment they sense a demand for electricity, they turn themselves on and start their inverting process. If there is no more demand, they go back to stand-by.

However, there are certain limitations to inverters. Usually they are rated by their maximum output. A

2,000 watt inverter puts out

2,000 watts at 110 volts using 18.181818 amps AC. That is probably enough to start and operate a circular saw or a hair dryer.

More powerful inverters put out up to 4,000 watts. And, of course, you can "stack" inverters to double either the voltage output or the watts they are producing. But something very fundamental has to be understood here: Let's assume we are using 2,000 watts at 110 volts and 18 amps AC. But our batteries deliver only 12 volts DC. This means that the electricity needed to create 2,000 watts at 110 volts and 18 amps has to be somewhat different when you start out with only 12 volts. So whatever goes into the inverter coming from the batteries on the DC side cannot be the same as what comes out on the AC side. If we remember, each time one function changes in the equation of voltage vs. amps or watts, another function will also

have to change.

Well, let's do the numbers, and convert from AC to DC and see how many amps we really use. Remember, the watts have to remain the same—since the device we are using makes this a condition and hence a fixed value. The volts change from 110 on the output side of the inverter to 12 volts on the input side. This means that the amps also will also have to change. The question is, if the 2,000 watts stay the same and the voltage changes from 110 to 12, how many amps do we need to deliver the same 2,000 watts?

Two thousand watts at 12 volts uses 166.66666 amps. Ooops! Well, that is a considerable difference. While our inverter puts out 18 amps, our batteries have to deliver 167 amps on the supply side of the inverter.

As a ball park figure in a 12 volt system, the batteries have to put out 10 times the amount of amps of the AC load. In a 24 volt system, the factor is five. If you had only a 220 amp-hr battery, how long could you run the above load? The answer is: 1.3 hours. And this is only in theory, not considering any other losses and the fact that you would not want to run your battery down that low. And, of course, don't forget Mr. Peukert's law.

So here lies the answer as to why we need a lot of batteries in order to dry our hair and why the cables between the batteries and the inverter have to be really big and the cables between the inverter and your hair dryer can be relatively small: 167 amps on one side and 18 amps on the other side. Big amps mean big cables because the electrons try to rush through the cable like our donkeys to the lake, all 167 at the same time to deliver the 2,000 watts. If the path, tunnel, or cable is too small, and the crowd still tries to fit through it all at once, some of them get rubbed against the tunnel

wall. As a result the tunnel (cable) gets pretty hot and, in extreme cases, will melt down.

Now, we know what a meltdown means—no, not the one in Chernobyl, but it still can cause a nice little fire.

So caution has to be taken when designing your system and choosing the right cable for every connection. (I am sorry if I repeat myself, but this point cannot be stressed often enough.)

Something else needs to be understood about inverters in particular and AC in general. AC means alternating current.

Yes, it alternates between plus and minus. The reason it does this is that it is created by a generator which is round. Remember those magnets and coils. (Never mind if you forgot, this will not be part of your final exam.) But whatever collects the electricity, the collector, is rotating inside the generator and alternately breaking the magnetic north and south pole—the one inside the magnetic field of the generator—or the positive and negative pole. Doing this over and over again (which is what happens if you run in circles), it creates cycles alternating between plus and minus, and if you would see it elongated in time or on the screen of a scope you would see what is called a *sine wave*.

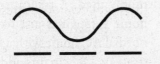

Up and down, up and down. If you count the amount of ups and downs as cycles per minute in our American 110 volt AC system you would count 60 cycles per minute. If you are a traveler and have been to Europe (and brought your scope with you), you would find that the Europeans use a different voltage (220 volts) and less cycles per minute (50).

Now, why is all this important with regard to our inverter? Well, nothing rotates inside our inverter. The electronic wizards outsmarted the simple mechanical rotation to create alternating current and came up not quite square and a little bit short. So the first inverters came out with what is called a *square sine wave*. The wave would look on a scope like a stairway to heaven and hell, or the rise and fall of the Roman Empire. The problem with it was that electronic equipment had a hard time dealing with it and so did electrical motors. TV screens had wavy lines across them, the power supplies of

73

computers burned out, and motors didn't know whether to run forward or in reverse.

So the electrical wizards started to modify the square wave so that it looked like a bumpy dirt road to heaven and, for lack of a better term, called it *the modified square wave*. This works fine for most AC applications unless you are a fanatical radio listener—you know, one of those who listens to talk radio only. Because all you would hear is a big buzz on your AM frequencies because the electromagnetic field the inverter induces into the house wiring sounds just like talk radio, or rather is an AM frequency.

A similar buzz can also be heard on certain brands of stereo speakers and sensitive amplifiers. There were also early reports of lines across your TV screen which I could never verify. In all, under certain conditions you were not getting quite what you bargained for as far as your AC inverter is concerned. For that reason, the wizards kept working on that dirt road and finally paved it. The result they call *the true-sine-wave inverter*. (I don't want to bore you, but even those waves aren't perfectly round.) Certainly a bit more expensive, but these top-of-the-line inverters are becoming more and more popular.

In most modern solar homes, the inverter runs just about everything, except the three areas mentioned in the DC section. (Everything that can create a phantom load: water pumps, heating, and some lighting.) This is one of the reasons inverters have to be bigger today. While 10 years ago a 300 to 600 watt inverter was sufficient, today 2,400 to 4,000 watt inverters are most commonly used. A typical solar home gets wired just like a normal AC home with a few strategic DC circuits added. Of course, there are a few differences we will talk about in the chapter about the electrical code. (See **Chapter Eight**, p. 108.)

In this section we learned that, in today's solar home, the inverter becomes the heart of the system. With few exceptions, it gives us all the fruits of our desires and, depending how strong these desires are and how well off we are financially, it can cover our basic needs or give us a smooth and noise-free performance.

And even though we still don't know how an inverter works (except that by definition it inverts), we know that our habits or our sophisticated electronic equipment plays the decisive factor in whether we

buy a modified sine wave inverter or a true sine wave inverter. However, sometimes our budget is the decisive factor. And, as always,

the size or the power of the inverter has to fit the size of our battery bank, which, as we already know, has to fit the size of our solar array. Are you slowly starting to get the picture?

Now, one quick anecdote and a word of warning. On occasion, I have been called to service a system and have noted that the batteries were dry and empty. What happened was that some repairmen dropped by, plugged their electrical

equipment right into the AC outlets, and went about their business, not knowing that the well they were drawing from was rather limited. The owner told me that all the workmen asked was: "Do you have AC power?"

Yes, you do have AC power but not as an unlimited resource. Always check what the unsuspecting repairman wants to plug into your system.

The owner of a local solar store visited one of my installation sites. (No, he wasn't going to check on my installation.) He talked to the owner/builder while I was busy elsewhere. The owner wanted to run his table saw on his generator as I had advised him to do. My good friend interfered, telling the owner that he could run his table saw off his solar system. And so it happened. Three cloudy days later, I returned to the site to finish up some business and noticed that the batteries were awfully low. I checked with the owner and he said that he only ran his table saw occasionally. "How did you set your new brick floor?" I asked. "Oh well, of course I used a tile saw to cut the bricks. The whole 1,200 square feet in three days. Great, isn't it?" "Oh yes," I replied.

He certainly was ill-advised by my friend who only wanted to point out that his system certainly could handle a table saw and, for that matter, a tile saw as well. He only forgot to consider that using a tool of that size for several hours a day when the clouds cover the sky may not be within the capacity of the battery bank.

It is also pretty common that "solar construction crews" persuade the owner to install the solar system ahead of time so they can run their saws and drills off it and do not have to use a noisy generator. By the time the house is finished, so are the batteries. An inexpensive generator (a noisy one) costs between $350 and $550. A battery bank usually costs considerably more.

G. SOME TOOLS
a. Voltmeters

One big aspect of maintaining a solar system is monitoring it well. Even

with small systems, you need some basic information about what is going on in your system. You always need to know the voltage of your batteries. This is the most basic and most important information. Some of the better charge controllers and some inverters provide you with this information. It is also good to know how many amps you are charging with and how many amp-hrs you are using.

With a small system, you can probably get away with just monitoring the voltage and the incoming charge. A digital voltmeter would be the best tool to monitor the system voltage because you need to know whether your batteries show 12.0 or 12.8 volts, which means the difference between a less than half full battery and a full battery, respectively. However, a good analog voltmeter can serve you well if it has an easy-to-read scale.

b. Amp Meters

To measure the amps, an analog meter is sufficient. You want to know what your charging is at peak times and how much you charge at low times (early morning, late afternoon, or cloudy days, etc.) Besides the charge coming in from the array, there are other parts of your system you want to monitor. There is the DC-load, which is how many amps all your DC equipment uses. This is an interesting area to monitor. Turn on the 25 watt DC light bulb that you installed in your utility room, the one that illuminates the very amp meter you are monitoring. In a 12 volt system, you will observe 2 amps on the amp meter. Turn on all the DC lights in the house and you may observe near 20 amps. This is equal to the value of a normal AC circuit in an average AC-only house. Monitoring this will tell you that every single light will draw a considerable amount of energy out of your battery system.

Besides the academic value, knowing what you draw out of your batteries can be of considerable importance especially when you know how many amp-hrs you have left in your system. Having an amp meter installed on the charge side as well as on the load side is therefore not a bad idea.

c. Amp-hr Meters

The third information you want to know about is your amp-hr status, especially in light of what we said above. You already know everything about amp-hours.

Your batteries store them (or rather electricity that can give you amp-hrs.) Your array can give them to the batteries for safekeeping and, last but not least, you can use them to run your coffee grinder. So it would be good to know exactly how much you have left in your system. An amp-hr meter can provided you with this information. It will tell you how many amp-hrs you have used up and how many you have regained during the day. There are amp-hr meters that measure DC amps as well, but there are much bigger toys ahead.

d. Combination Monitoring Devices

There are several monitoring devices on the market—from simple to very sophisticated. The price range is also from simple (under $100) to sophisticated ($400). So make sure you understand what and how much you need to monitor.

Some of the more sophisticated meters can monitor almost all parts of your system, including incoming power, incoming voltage, battery voltage compensated for battery temperature, other incoming sources, amp-hrs used, amps in use, etc. Some even allow you to call this information up via computer modem,

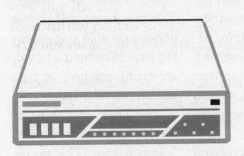

giving you a spreadsheet of your energy use for two whole weeks. You can tell exactly at what time on what day you went to bed or how long you (or your kids) watched TV. A real big brother watching you. It also gives your solar installer a chance to remotely troubleshoot your system by calling up information via modem.

e. Power Centers

Some companies offer complete DC or DC/AC-load-centers—better known as Power Centers—which you mount on the wall and then run all your cables into. These panels, expensive as they are, are very helpful as they include everything you need for your system as far as disconnects, fuses, breakers, charge controlling, and monitoring systems are concerned. They are also UL approved which makes inspection time easier. But you would only get your money's worth if you have a bigger system, as these devices range in price from $1,200 to $1,600.

These power centers have become very popular on medium- to large-sized systems. And, indeed, they usually include all you really need. Solar installers love them. But they are far from

79

what the manufacturers want you to believe, as far as their installation is concerned: "Just nail them to the wall, hook your wires into them, and be done." Not quite! I found that it still needs quite some expertise to run your wires into them. There are code requirements to be followed, wiring techniques to be observed, and a thorough understanding of its components is necessary to accomplish a safe installation. After, all you are dealing with 200 to 400 amps inside these boxes, which can give you quite a scare when the sparks start to fly.

I also ran into several problems as far as the designs were concerned, as well as faulty installations and arrangements inside the boxes. Again, this should not undermine the convenience of these load centers—once properly installed, they are the nicest invention after the the invention of the wheel. However, on smaller systems, you may assemble the components yourself for a fraction of the price of those panels.

f. Other Components
1. Converters

Other helpful components for your system may be *converters*. We have heard of them earlier, have we not? Again, a converter transforms a DC voltage to a higher or lower DC voltage. This may be needed if you decided on a 24 volt system but want to connect a 12 volt answering machine or tape recorder to your DC.

Their cost starts at about $60 for a very limited 2 amp output. This means you can use them only for small electronic gadgets. These smaller converters usually get wired into an outlet box and include a cigarette lighter receptacle. The advantage to this is that you can install them at any point in your 24 volt DC line and have 12 volts right there while everywhere else along this line you still have 24 volts. Plan on installing a larger than normal box to receive this device, because it puts out a lot of heat which has to be able to dissipate— or it simply shuts down on you.

Of course, the cigarette lighter receptacle may prove a problem when the inspector browses through the house since they are not UL listed.

If you need more power than just 2 amps, you will have to install a dedicated 12 volt line and step your 24 volt DC down to 12 volts right at the beginning. If you want to connect some 12 volt halogen lights to a 24 volt system, you may consider wiring them in series— remember, two 12 volt lights in series gives you one 24 volt light.

2. Transformers

Transformers are used more frequently since bigger systems give you more power which results in new demands. They transform an AC current to another higher or lower AC current. If you have a deep well and need a powerful well pump, you most likely need 220 volts. You can achieve this by either "stacking" two inverters and having 110 volts times two or installing a transformer in line with a single (and powerful) inverter.

A 15 amp transformer which can run a water pump up to 1 1/2 hp costs around $400 to $500, which still is only a fraction of what a second inverter would cost.

3. Battery Chargers

Yes, your batteries get charged by the sun, that's why it's called a solar system. But just like our minds, sometimes all seems to be a little cloudy. And if the cloudiness prevails for some days, you start running low. Of course, there is always "Prozac," but try feeding it to your batteries!

Now you start conserving energy. No more

soaps, the kids have to learn how to read books again, even the old Scrabble game gets revived, you start conversations again, talking and listening to each other. Of

course, all seems a bit strange without the constant squeal of cartoons in the background. Hey, that sounds like Life again—why would you want to change anything? (Oops, my editor will say, "we can't print this, this is not politically correct. People who are interested in photovoltaic systems do not behave like this." I tell her that I am just trying to entertain, but she does not want to hear it. "I know," I tell her, "this is supposed to be educational, but why does education have to be dull?")

Enough—let's just assume that, due to a stationary warm front or maybe an occluded front, the cloud layer prevails for several days, which causes the voltage in your system—due to normal use—to reach a point where the protective devices built into it to alert you to a low voltage situation and eventually shut down certain components like the inverter. Now what? The forecast says the front may linger for another week or so. You have to recharge your system from an alternate source. If you have hydropower, the rain coming out of these clouds may speed up your turbine. If you have wind power— well, a stationary front may not create much wind.

But you planned ahead and bought a generator. Now, the same is true for solar batteries as it is for your car battery. You need a charger to recharge them. If you have a small battery charger which you occasionally use for your car battery, it most likely will not work for your battery bank unless you either have a very powerful charger or a very small battery bank. The average car battery charger provides you with 10 to 20 amps of charging capacity. This may not be enough for your 800 amp-hr battery bank because you would have to run your generator for how long to recharge the batteries?

Let's say you are down to rock bottom, which is 20% left in your batteries. You have 160 amp-hrs left. Just to bring them back up to half capacity, which is back to 400 amp-hrs, will take 12 hours (you need to recharge 240 amp-hrs, divided by 20 amps = 12 hrs). Running your generator for that long does not seem to be a good idea. You are wasting fuel, polluting the environment, and wearing down your generator. And we are not considering other things, like using energy while you charge, your charger not delivering full capacity, etc. You want to recharge your battery bank as fast as possible. (This only applies to lead-acid batteries. Nickel cadmium and nickel iron batteries do not like shock treatments like this.)

Fortunately, there are powerful battery chargers available. They can recharge your batteries with up to 190 amps. Unfortunately, they are very costly, ranging from $200 for a 30 amp charger to $2,300 for a 190 amp charger. Of course, you want to size the charger according to the size of your battery bank. If you would shoot at your small 440 amp-hr bank with 190 amps, you would literally blow it up.

Fortunately, most modern inverters come with integrated battery chargers for little extra money. They usually have what is called a *transfer switch* which, upon sensing an incoming voltage, automatically changes over to the charging mode and sends whatever energy is not used for other purposes (soap operas) to the battery bank. You may ask: If they come with a charger built in, then why make so much fuss? The answer is, education.

4. Generators

What do we need to know about generators? The answer is, more than you possibly would want to. You need to know how to do an oil change, how to change the spark plugs, how to fuel them, etc. Because you most likely do not want to take them to a service dealer every time the fuel filter is plugged up. They are just too heavy.

But this is not what we want to talk about here. What size and type of generator do we need for our system? Let's talk about the type first. If you look at the different generators on the market, you may notice that they differ in price considerably. For a 4,000 watt generator, you can pay as little as $500 and as much as $3,500. How come?

A light inexpensive generator is for occasional use only. These light generators are usually of the pull-string-starting type. They are very noisy. They are very, very noisy! All connections are usually of the plug-in type. If you service them properly, and if you can live with the noise, they can last you a while. (I still use one of them. I built two houses with it and still charge a small system with it occasionally. It is now 6 years old.)

Heavy-duty generators, of course, last much longer. Their noise level is considerably lower. They often come with features like battery starting and remote starting, which means you can even program your inverter to start the generator in a low voltage situation. Some brands can also be powered with propane gas which is environmentally friendlier. Again, it is a matter of preference and a matter of budget.

But it is also a matter of power. What size generator do you need? As mentioned earlier, if you want to use a battery charger that can provide 190 amps, how big would your generator have to be? (Remember: amps x volts = watts.) Well, about 20,000 watts or 20 kilowatts. Your little 4,000 watt inverter would break its back to provide such a load. But let's be realistic! Most inverters come with battery chargers up to 70 amps, which means you still need over 7,000 watts to charge the batteries. But usually you will not need a full charge, because your batteries are not that low and the battery charger senses the appropriate charge capacity needed and adjusts accordingly. Similar to car battery chargers, you will notice that the charge amps go down as the batteries fill up.

Another consideration, of course, is the size of your system. If you have a 4,000 watt inverter, you are able to use up to 36 amps in your house. If your genera-

tor is a back-up generator, which means it is not only designated to recharge your batteries but also to supply the rest of the house with power, you should have at least 36 amps available for your needs in your home. And certainly you need some extra to recharge your battery bank. Ideally this would mean at least a 6,000 watt generator. Of course, a generator of this size is very expensive. But if you consider an automatic generator starting feature, you will have to size the generator this big.

Here is why: If you run a big load off your batteries for quite some time (let's say a total of 30 amps), the batteries reach a low voltage stage. This in effect would activate the automatic generator-starting feature. The generator would start and would have to be able to cope with the immediate power demands (a 30 amp load plus the low battery condition with a high draw from your charger). This may strain your generator to such an extent that it would have to shut down and you would get nothing.

Of course, if you have to start your generator manually, you can adjust your load and shut down heavy equipment. Now without your back hoe and steamroller and German WW2 tank running, you can attend to the needs of your battery bank.

Here is another consideration that you may not be aware of. If you live in the Southwest with its simplistic furniture and rugged lifestyle, you may know that the average elevation is around 5,000 feet MSL (Mean Sea Level). Now, every pilot of small airplanes knows that performance goes to pot with rising altitude. (Outdoor enthusiasts know that this is not only true for combustion engines but also for the human engine.) Above 5,000 feet you can count on engine performance to be reduced by about 25%—which means you get only 75% of your horsepower or watts. From a 6,000 watt generator, you would get about than 4,500 watts. This puts your choice of a generator into the next higher category.

The performance of the combustion engine suffers in higher elevations due to lack of oxygen, which it needs to burn fuel. This lack of oxygen does not affect electric motors—they pull the same old amps that they pulled at sea level. So the *demand* on the generator remains the same but the *performance* of the generator suffers with altitude.

In all, choosing the right generator depends on your willingness to either participate in its supervised use or your willingness to come up with the cash to afford the luxury of an automated system. If you only use it to recharge your batteries and run an occasional power tool on it, you can get away with a simple and small generator. If you want to have a full back-up system, you need to consider the full capacity of your electrical system plus the charging power of the battery charger. In either case, as pilots would say, watch your altitude.

There are many more accessories available. In **Chapter Ten**, we will talk more about tips and gadgets, what to buy and what not to buy. (See p. 127.)

In this chapter we talked about accessories most commonly used in solar systems. They are extra tools to monitor, simplify, and customize your system. As we learned about the components of your system, we saw that every component ties into the next. From the tree-top to the deep soil, they need to be sized to consider all the factors involved. If you change one part, all the others may have to change. It all is based on your needs and/or your budget. Sizing a system is a tricky undertaking and eventually needs to be done by an expert.

In the next chapter we will talk about an easy way to do a preliminary sizing, just to see what we are up against. (**Chapter Seven**, p. 88.) Some people need to outweigh the cost of bringing grid power to their house against the cost of a solar system. Understanding what is involved and how a system works will help you make such a decision.

Chapter Seven
SIZING THE SYSTEM
How Big a Garden?

All of the previous information (**Chapters One-Six**) is not necessary to understand this chapter, but it surely will make things a lot easier. When I went to Hippy Alfred to size my system, I was gullible, and he could have sold me anything. Fortunately I could not afford it, and that is the very first factor in sizing your system. It is not just a question of how big a garden you want, but how big a garden you can afford.

If your budget is as limited as our country's and you do not want to create a deficit which you then have to balance over seven years or more, you may look into which parts of your system can be expanded later and which parts cannot.

Alfred (I will leave

his title out for now) had me list the electrical appliances I wanted to use. Not knowing exactly why, I listed my TV, stereo, coffee grinder, dishwasher, washing machine, dryer, computer, printer, electric saws, electric tooth brush, microwave, hair dryer, etc. Then he did the calculations and came up with a size and price for a system (including tax) that

one thing: Write down everything you might possibly want in your solar home. Then begin the process of elimination. First, take off everything that can be run on gas. (If this includes a dryer, a fridge, a radiant heat boiler, and an electronic oven/range, remember that parts of these appliances still need electric power to operate.)

would have made Donald Trump cough.

But knowing how things work can help us to choose the appropriate system. Alfred was right about

Since we know that every heating or cooling appliance needs an enormous amount of power, you can discard the idea of running them on solar electric alone. This is true too of sophisticated printers (laser printers) for your computer. A hair dryer, if used only for a few seconds, is probably acceptable. DC fridges are designed to use very little, but still demand quite some additional, power. Also big electric tools, i.e., electric motors, fall into this category, not because they use very much while running, but because they use a lot of amps for starting or when getting stuck while cutting through some wood.

This is particularly important if you have a deep well and need a well pump. Most common well pumps operate on 220 volts, as your well driller will tell you. But there are some 12 volt or 24 volt to 60 volt DC models available which can pump as deep as 600 feet and may do the job just fine, except that your well driller does not know about them and, most importantly, does not want to know about them.

You will find that only a very few plumbers, well drillers, and electricians are knowledgeable about solar applications. The others, rather than admit their ignorance, will try to put you down and make you feel stupid. One reason for this is that they are too lazy to learn about solar water/electric—plus getting involved with something that is nonstandard, and therefore has to be custom designed, makes them uneasy.

Here is a little story to illustrate the subject. Someone had bought a solar home. A shallow well (120 feet) came with it. To supply him with water, I installed a DC pump which resulted in a gain of 3/4 gallon per minute to be pumped into a holding tank level with the house. Also with the house came a DC booster pump and a pressure tank to supply the house with proper water pressure. However, the pressure pump was not yet installed. (It was lying right next to the pressure tank.) The plumber contracted by the new owner ran all the water lines from the well to the holding tank and to the pressure tank. When I was called to do the power connections, I noticed that the booster pump was not in-

him, the plumber, he did a neat job of pressurizing the water system.

I don't mean to be arrogant in telling this story. It is a pretty common occurrence that contractors and journeymen are very set in their local customary ways and show little flexibility when it comes to alternative ways of doing things.

stalled. I asked the plumber how he thought he was going to create water pressure in the system. He said he had not thought about it because usually the well pump creates the pressure. Then I asked him what he thought the holding tank was used for, which was actually slightly lower than the level of the house. He said he had not thought about it. I asked him what he thought the booster pump was for, which was lying right next to the pressure tank. Well, guess what? He had not thought about it. After I, the electrician, explained everything to the

They also fear the national and state codes for solar applications, which are difficult to interpret. My experience is that, usually, the inspectors also have very little to say and keep their inspections within the area of their expertise. But beware if they know something, or feel they should know something, or think they do know

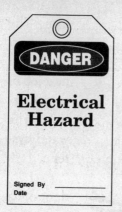

DANGER

Electrical Hazard

Signed By _____
Date _____

some-thing. I will dedi-cate a whole chapter (*Chapter Eight*, p. 108) to the National Electric Code and inspections. But do not worry, it will mostly be anecdotal—except that I will explain the essence of the NEC to you. (You may want to keep your code book handy.)

"You can't always get what you want. . . ."

In the end, it is not what you want but what you need. So, let's discuss the voltage you might need. Batteries can be wired in all kinds of configurations. Usually in solar applications, you will find 12 volt systems and 24 volt systems.

"*There's a war between the left and right, a war between the black and white, a war between the odd and the even. . .*"

There is also a war going on between the propo-nents of 12 volt and 24 volt systems. It often depends on which one you run into as to what system you will end up with, unless you know what they are talking about. Usu-ally the argument goes like this:

"Uh, for 12 volt sys-tems you need much bigger cables than for 24 volt sys-tems. Cost you a lot more money."

"Yeah, but you can get more appliances and light fixtures for 12 volt systems because the whole RV world is full of them."

"But 24 volts is the only way to go when your system is big and you can al-ways convert it down to 12 volts in certain locations."

"But why waste $60 for every converter when you can have it right from the source?"

As you may guess,

this discussion could fill a whole chapter and you still would not be any smarter. To illustrate this, here is a short story:

I was called by a woman who had bought a small system and wanted me to install it. Her objectives were to have a few lights AC as well as DC, a few receptacles and, most important, a cordless phone/answering machine and her CD player both running on DC. She also mentioned that she wanted some provision for a future expansion to a small workshop running a few small electrical motors.

Based on her last statement, the solar supplier had sold her equipment based on 24 volt DC (to cater to her future need to run some small grinders and an electric sewing machine).

To supply her DC needs, he sold her a converter, the one mentioned earlier, which has to be installed inside an electrical junction box. It supplied her with 12 volt DC and 2 amps.

After everything was installed, her installer drove happily into the setting sun only to find her lovely voice on his answering machine at home. "My CD player kicks in and out, it buzzes and my telephone doesn't work at all any more." The installer rushed back the next day and tested everything. And everything worked well. "If it ain't broke, I can't fix it," he said and left.

Now you can imagine what happened next. After the converter was exchanged three times and an extra large box was installed around it with extra holes for ventilation, it finally ran the CD player and the phone, but it still buzzed. Nobody could solve this puzzle. And yes, according to the name plates on the devices, there should have been no such problem. The manufacturer of the converter could not explain the buzz away either. And so, dedicated to her music, she ran her CD player on 110 volts (it didn't buzz there, in spite of her inverter delivering a modified sine wave) and unplugged it each time she stopped using it to prevent a phantom load.

What is the moral of the story? If she had gotten a 12 volt system in the first place, all this trouble would have been avoided. Years later, she still has not built her work shop, which was the reason for the supplier to sell her a 24 volt system. She called me recently to tell me that her love for music had forced her to buy a 200 watt stereo system, and she would be willing to expand her system to be able to listen to music all day long.

One can easily fill a page or two with these kinds of stories and the above arguments but let's look at a few facts. The size of the wire is of no importance except in two instances. In the internal house wiring for DC, you want the biggest wire you can possibly manage to compensate for losses, no matter whether you run 12 volts or 24 volts through them. This is usually a number 10 cable. And it may be true that you need a thicker cable for your electric DC fridge.

It also makes a difference when it comes to the line between the array and the control panel and the same is true for the cable between the batteries and the inverter. Here you can run quickly into big or very big wire sizes if the distance between the array and the house is considerable. This may also amount to a very big expense.

I had a customer whose fairly big 16-panel array was mounted about 500 feet away from the house. Considering the amps travelling this distance, I had to run #2 O/D cable, which is thicker than a thumb. It also cost her about $1,000 just for the cable.

But these are exceptions and need to be dealt with as such. Generally, for small- and medium-sized systems, 12 volts is acceptable.

Also considering the price of wire vs. the expense of solar panels, controllers, and inverters, wire is still pretty cheap. And why would you pay all this money for your equipment and then start being miserly and lose power due to having wires which are too small? You probably want to oversize anyway, in case you ever need to upgrade your system. As an example, a roll of 250 feet of 12-2 wire costs about $35 while a roll of 10-2 runs $50 (a 6 cent per foot difference).

Generally you will see bigger systems laid out in 24 volts. But it's not the size of the system that determines the voltage but, again, what you plan on doing with it. If you have special needs such as a deep well pump which you want to run on DC, you will have to choose 24 volts because these pumps only come in 24 volts or higher (and in this case the cable length of several hundred feet does favor 24 volts, which will have less voltage drop over this distance). But if you do not need any of this special equipment and everything is located in close proximity to the house and the DC control panel, a 12 volt system will serve you just fine.

Systems are getting bigger and bigger. (And you thought this was only a problem of the human anatomy, right?) With increased power demands, you want to get as much power for your money as you can. So the latest trend is a doubling of voltage to 48 volts for very big systems. One reason for this is that equipment gets more efficient and a little cheaper in this voltage range, simply because components inside the equipment can be smaller at this higher voltage. One example is a popular inverter which delivers 4,000 watts in 12 volts as well as in 24 volts but over 5,000 watts as a 48 volt model.

Forty-eight volts is most likely the limit for home power systems because the NEC, the National Electrical Code, requires drastic measures for installing anything over 50 volts DC. (If you are standing barefoot in puddle and put your hands on a 50 volt line, you can feel a nice buzz.) The open array current on a 48 volt system can be as high as 70 volts, which puts the system into a different category as far as the electrical code is concerned.

But back to our DC application. When it comes to booster pumps, for example, there are some very inexpensive RV pumps available which run on 12 volts, which work very well to pressurize small- and medium-sized water systems. It is true that most DC appliances and light fixtures come from the RV world and hence are much cheaper, due to mass production, than what is available in 24 volts (or 48 volts).

Now here is the first decision you will have to make. If you plan on hanging a lot of 12 volt lights (like halogens which are commonly available in 12 volts), the pendulum swings to 12 volts. If you need a specific pump which is only available in 24 volts, you may choose the 24 volt system. But before you make a decision, educate yourself or get a second opinion. Most of the solar suppliers are very helpful on the telephone, some even have technical hotlines where you can talk directly to some technician about your situation.

Your local solar dealer may advise you one way, but if you can discuss things with him or her, or at least ask the right questions, you may end up not only with what you need, but you will also know why.

All this may take a little bit of effort but in the long run you will be better off because you have to live in your house, not your consultants, dealers, or architects. (You will hear me say this a lot, because it is true.)

The Load Calculation

Let's get back to the calculations that Alfred did for me. (Remember, the one he used a pressure-treated 2x4 on?) Well, this is a method you may use for the final planning stage and its results are enlightening only for engineering purposes.

What Alfred had me do was what I call the guessing game. "Do you watch TV?" he asked, and I had to admit I did. So I gave him an approximate amount of time that I watch TV per day, of course deceiving him and myself with an understatement. This is the procedure you use to find out what your total load will be. You answer questions about how long you will dry your hair, how often you wash laundry, how much time you spend on the Internet, etc. The end result of all your consumption is what your daily use will be in either watts or amps. Now you can imagine how many mistakes this procedure can contain, especially when you know that every extra hour you use electricity can cost you x-amount of dollars in equipment. Wishful thinking will sneak into this equation in no time.

My experience in the field is that almost 50% of the systems I install or service are undersized. (One could almost specialize in upgrading systems only.)

Here is a formula which I call the *Adi-factor*, to find out how well you know yourself. Take an estimate about next month's long distance phone bill. Write it down and compare it to reality a month later. The difference between your estimate and the actual bill will give you a percentage that tells you how well you know your habits. You can do the same with your electric or gas bill.

Now back to your load calculations (also known as the collection of assumptions and self-deceptions. Or the presidential factor: Living in denial!). Eventually you will have to do one because you need a figure to start the calculations from.

The best way of doing this is to use one of the forms provided in almost every solar mail order catalogue (or supplied in the back of this book [see *Appendix E*, p. 206]). In essence, you try to find out how many watts you are using per day. This figure will be adjusted for losses due to inefficiencies of the inverter or the wiring, mismatching of panels and batteries, etc., to come to the total needed watts. Now you know how much elecricity you need to meet your needs.

Next you find the right amount of panels with the right wattage to supply this amount. You first divide your total watts needed by the average time the sun is charging your array. If you have a standard array (a fixed one), this is somewhere between five and seven hours. Now you have a figure that represents the size of your array. In other words, if you need a total of 3,000 watts per day and you charge for 6 hours a day, you need an array that can produce 500 watts. (You divided 3,000 by 6 hours.)

Then you need to find out how many panels can make up such an array. If you buy 50 watt panels, for example, you need 10.

The next thing you

need to find out is the right amount and size of batteries to store this power for as many days as you can afford and, *viola* (pardon my French), here is the size of your system.

Usually, you want a reserve of power, so as not to run down your batteries every day and also in case the sun doesn't shine for some time. This reserve can be calculated according to your budget. Mostly it is somewhere between three to five days.

Let's use our example of needing 3,000 watts and an array of 500 watts to provide the energy. If you would store one day's energy, you would multiply your 3,000 watts by 1. OK, that was easy. If you need three days, you multiply it by 3. Now you have 9,000 watts stored in your batteries. But, remember that deep cycle batteries should only be discharged by 80%, so 20% is unusable energy. We need to increase our battery bank to accommodate this 20%. So we multiply the 9,000 watts by 1.2 to increase the size of the battery bank by 20% and arrive at 10,800 watts. This is how big our battery bank needs to be to store our daily use of 3,000 watts for three days, considering our deep cycle batteries can only be discharged up to 80%.

OK, how many batteries do we need? Oh well, there is lots of confusion here, I can tell you! Batteries usually come in 6 volts and a certain amount of amp-hrs. The pretty common golf cart battery has a capacity of 220 amp-hrs at 6 volts.

Do we still remember the one and only formula: **Volt x Amps = Watts**? Yes, we do! So let's fill in our values so we can determine how many watts a battery has: 6 volts x 220 amps = 1,320 watts. That's how much one battery can give up. So let's divide our 10,800 watts, the total we need for three days, by 1,320 and we get 8.18 batteries at 6 volts. You would need a bank of eight batteries to store this power for three days.

Initially, however, you want a ballpark figure to see what size system would roughly fit your needs and here is a much simpler method which is good for the pre-planning stage and usually holds true even when it comes down to the nitty gritty.

Look at the biggest energy consumer you want to run off your system on a regular basis. If you are a computer nerd and have to print a lot, see what printer is available with the best results and the lowest power demand. Now, this is your main load. Now you ask yourself, is there any other load that may interfere with it? Is your washing machine going to run while you need to print? Do you have to run any heavy duty motors during that time? If you choose

to live in the outback, the downunder, you probably can wait to shred your wheat until your printer has stopped. As the saying goes: living in a solar house requires a little bit of planning. If you can't do that and/or cannot afford a big enough system, you may reconsider whether a nice little apartment on 5th Avenue would

not be better suited to your lifestyle.

After you have established the biggest user in watts, look for an inverter that comfortably provides you with this wattage. This is the core of your system, your main user. Now you need to plan around this inverter. You have to back it up with enough battery power to run your printer for the time needed and you have to size your array to give your batteries enough charging power.

How do we do this? Well, let's do the numbers again. I know you don't like it, so let's keep it simple. (My publisher said: "This book is not for Dummies, we cater to educated people who are capable of learning." The response of some of my readers to this was: "When it comes to photovoltaic I wish you would write for idiots." Now what?) Well, you figured out that you need a 2,000 watt inverter, because you are going to use your hair dryer on a regular basis (I am not implying anything here) which needs 1,200 watts. This is your biggest user, plus a reserve in case your TV or some lights are on as well. You will have to size your battery bank accordingly, which means you have to back your inverter up with enough batteries so that your hair dryer or an equivalent load can run for an adequate amount of time.

Power times *time* gives you the magic numbers. You are giving hair cuts in your home and need to use your hair dryer for six minutes after each cut. You average 10 cuts a day—that makes for 60 minutes at 1,200 watts. You get 1 hour times 1,200 watts which is 1,200 watt-hrs. Easy so far?

Twelve hundred watt-hrs divided by 12 volts gives you 100 amp-hrs. This is what you use just for your profession. Add a few hours of TV time, lights, etc., and you end up with another 100 amp-hrs. Therefore, 200 amp-hrs per day will be used and will need to be replaced. Now count on several cloudy days (depending on your region)—let's say about three days—and you are up to 600 amp-hrs. Of course, you do not want to use more than 80% of your deep cycle batteries, so add another 200 amp-hrs for good measure. You need a battery bank capable of delivering 800 amp-hrs.

Now, about the charging process. Replacing 200 amp-hrs takes how many panels

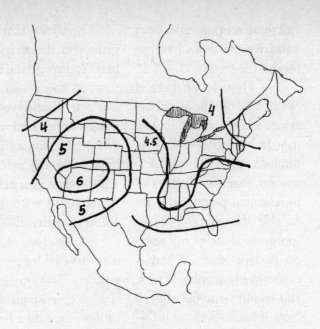

at 50 watts each in a 12 volt system if you have an average of five hours of sunlight per day? How do I know what my average hours of sunlight are in my area? You look it up in a chart that you can find in most solar catalogs or on this page.

Now, you do not have to do the above calculations, but here they are for the curious: 200 amp-hrs are 1,200 watt-hrs. (Remember the one and only magic formula: **Volts x Amps =**

Watts.) One panel at 50 watts times 5 hours gives you 250 watt-hrs; 1,200 divided by 250 gives you about five panels. Just to make things even, usually you fit six or eight panels on a rack, so call it six panels. The same is true for the battery bank. If you base your calculations on 6 volt and 220 amp-hr batteries, you may come up with eight batteries in a 12 volt system, which will give you 880 amp-hrs.

After struggling with

all these numbers, let me tell you, there is a much simpler way of doing it. (I know it is mean to tell you this after all the brain work.) There are three basic sizes of systems: small, medium, and large. What works for your dresses and shirts works for your power as well.

A small system, for a cabin with only a few lights and some demand for TV, radio, etc., needs a small inverter, 400 to 1,000 watts. It needs between four and six batteries at 220 amp-hrs and 6 volts each. It needs to recharge power from your array at 100 to 200 watts.

A medium system needs an inverter with 1,200 to 2,400 watts. It needs about eight to 12 batteries at 220 amp-hrs each and 200 to 400 watts of panels.

A big system, containing a big water pump, and provisions for bigger power tools and extended use of computers and printers, needs an inverter with 2,400 to 4,000 watts. We are talking here of 600 to 1,200 watts of charging power. As far as the batteries are concerned, we have to go up in size. Interconnecting more than 12 batteries in the 220 amp-hrs category is not advisable. There are too many things that can go wrong in a big battery bank like this. Loose jumper cables can create arcing or slowing down of the flow of electricity between batteries, one cell can go bad and take the whole system down, there can be corrosion at terminals, low water levels, etc. It is best to find bigger and fewer cells, like 360 amp-hr, 6 volt batteries or powerful 2 volt industrial cells which are used for fork lifts (not for golf carts!).

Just find yourself within one of these categories and you have a good start as far as your budget is concerned.

Upgrading

If you need a bigger system but cannot afford it

right away, you need to plan a smaller system so it can be upgraded. Now, which components can be upgraded and which cannot? Let's go from top to bottom:

1. *The array.* You can always add and mix panels of different design, size, and wattage as long as you end up with the same or a similar voltage. If you have some older recycled panels which give you 100 watts at 24 volts open circuit and you buy two panels delivering 75 watts at 20 volts open circuit, can you mix them? Yes, you can! You connect the two new panels in parallel (plus to plus and minus to minus) and combine them with the plus and minus of your old panels (provided, of course, that your wire between the array and the control panel was sized big enough). This will give you an open circuit voltage sufficient to charge a 12 volt system. So there are no problems adding panels to your system.

2. *Charge controller.*

You need to know now how many total amps are coming down the line—not only for the size of your cable but also for the size of the charge controller. Your old panels gave you about 8 amps charge power. The new panels add another 12 amps. Can your charge controller handle 20 amps? If not, you may have to upgrade it.

3. *The batteries.* The biggest problem when it comes to upgrading are the batteries. Unless you upgrade within a few months, batteries do not like "new blood" among them. Batteries age rapidly and develop habits, just like people. If you add new batteries to old ones, the old ones will pull the new ones down down to their level. The new ones will perform to the standards of the old ones, which is a waste. So either plan on a slightly bigger bank initially, for future purposes, or upgrade your system when the batteries are new.

Of course you can

combine new panels with new batteries and have a separate system for separate needs. I have seen systems with 12 volts for lighting and 24 volts for pumping and refrigeration.

4. *Inverter.* If you increase your power needs, you most likely also want to have a bigger inverter, unless your power needs are on the DC side (adding an electric fridge, a pump, etc.). If you try to sell your used car, you might get a fair price for it. If you try to sell your used batteries, you may get the recycling value for them. Inverters are like used cars, they keep a fair market value and are good trade-ins when you need to upgrade to a bigger one. Some inverters can be modified for a higher wattage, and some inverters can be interconnected with another one of the same kind to double their capacity.

Conclusion

This chapter is one of the most important chapters in this book because it deals directly with your present and future cash flow. It also deals with your future lifestyle and may affect your relationship with your house mates, including but not limited to wife, husband, kids, and pets.

So it is very important that you understand this chapter well. Needless to say, if you want to understand this chapter well, you have to understand the other chapters as well. (Tricky, isn't it?)

I just paid a visit to an old customer who read the first edition of this book, the shorter and funnier version. In addition to the fact that she regretted hearing that I had taken some of the "wisecracks" out and added a lot more material ("The shorter, simpler, and funnier, the better," she said), she had problems understanding the chapter with the dog sled. I told her, I am no expert on dog sleds. I just know now that dogs, like donkeys, are paired up—but she interrupted and pointed out that even the simplest of formulas makes her dizzy. Short of rewriting the whole chapter, I want to explain again that formulas are like real life. If you change one thing, all the others change as well. If you just bought a new 4x4 and want to put it to a test, you

can drive up the steepest hill, but only if you put it in the right gear. To achieve the full horsepower (watts), you need to change to a lower gear ratio (volts) by shifting into first or second gear. What you will notice is that the engine has to work harder to produce a higher RPM (amps). For some strange reason, she could relate to that much better. If I don't have you fully confused by now, just wait until the

last chapter. I am sure I will have succeeded by then.

Let's see what we can remember about this chapter. We started out by asking what size garden you want and what size you can afford. Good ole Alfred, who should get all the credit for this book since, without his negative support, I would not have gone through all this trouble, made me do what every house builder does: first find out what you really want.

We need to do that too, even if only to learn that you can't always get what you want. It is very hard at first for the non-technically inclined person to see why such a small item as a hair dryer needs 10 times as much power as a big 25 inch TV.

Going through the list of "NO NOs" for a solar system will soon make it clear whether you want to live this way or not. The list is short—it's called the *ABCs of NO-NOs:*

A. Everything that creates heat or cold (with the exception of a DC fridge).

B. Everything that has to be on continuously and needs AC-power, hence creating what is called a phantom load (unless you can afford a big enough system to support your or your kids' TV habits, in which case it is called a load of soap).

C. Everything that requires more than 110 volts AC (except maybe if you need a certain water pump or stack two inverters to give you a 220 volt set-up).

All this we learned in this chapter, did we not? But we also learned that we can size our system according to our needs or budget. We can plan it in such a way that we can upgrade its components even if we have to trade in or sell some of them.

We learned that batteries age like people, but unlike people, a young addition will lose its vitality fast when surrounded by old-timers. We also learned that panels can be mixed and inverters can be upgraded or interconnected. But most of all we learned that you can lose your brain in endless calculations in order to find just the right size of system and then find out that you cannot afford it.

But for initial investigation, you can size systems into three categories: SM, M, and L (for the shamelessly rich, one might add an XL).

We also learned that there is a war out there between the 12-volties and the 24-volties, and soon to be the 48ters. None of them is right of course. As always, it all depends on your needs. And last but not least, we learned that even after we did our homework thoroughly, we still may need a second opinion on certain aspects because, in the end, we may hear ourselves saying, "next time I would do it all differently."

Chapter Eight
TO CODE OR NOT TO CODE
The Inspector Is God

I know, this is future music to you if you are planing on buying a system or a solar home, or are just interested in photovoltaic in general. But a little bit of inspection blues can't hurt nobody!

In the (good) old days when photovoltaic was reserved for the brave and daring, nobody cared for codes of any kind. Everybody was a pioneer, a test pilot, and, if the system didn't blow up, it was already a big success. Unfortunately, sometimes it did blow up and, as the outbacks slowly moved closer to civilization, it became a concern of "The Authority Having Jurisdiction." (If you consider that a small golf cart battery when short circuited can spark with the force of close to 8,000 amps, you also may become concerned.)

So in 1984, photovoltaic was formally addressed in the NEC, the *National Electrical Code*. It may be of interest to know that the original NEC document was developed in 1897.

The purpose of the NEC (besides harassing electricians) is to safeguard persons from electrical shocks and properties from electrical fires. The majority of early solar communities, as described above, strongly believed in the alternative lifestyle and rejected any involvement and control by "Authorities Having Jurisdiction." The strong belief in the do-it-yourself way of life also prevented educational intervention by trained personnel. As a result, most systems installed in those days would scare the living daylights out of any qualified person familiar with today's interpretation of the NEC. A lot of today's solar installers were yesterday's pioneers and still reject the notion of an electrical code interfering with their low-voltage installations. Some also think that low voltage—as described in the NEC: a voltage lower than 50 volts—is not covered by the code. Wrong!

What they seem to forget is that things have changed. Those systems in the beginning days could be one solar panel feeding some old truck batteries dangling at the end of a piece of

Romex wire. Today's systems have battery banks that can create short circuit currents of 100,000 amps and solar arrays with short circuit currents of 60 amps or more.

Of course, having a copy of the NEC, reading it, and understanding it are three entirely different things. Myself, being of "unnatural" origin in this country with English being my second language, I have mastered to some degree not only the immigration law

but also—being a pilot—the aviation regulations. But nothing compares to the language and wording in the NEC. It leaves room for interpretation and clarification that would plug up a black hole in no time. (That's what every Inspector builds his/

her empire on.)

The NEC covers the electrical field with 800 articles over 1,000 pages. Article 690 covers Solar Photovoltaic Systems in 74 paragraphs over 10 pages. Of course, the trick to covering such a vast subject in so few pages is well-known and used in all law books. It is called "referrals." Exactly 19 times the article refers to other articles in the NEC, most of them covering the principles of general wiring techniques, grounding systems, fusing, etc. In other words, you have to be an electrician to understand the Code and even this may not help you.

At times, the Code is very clear and distinct: "If the equipment is energized from more than one source, the disconnection means shall be grouped and identified." (690-15) But you also have to know that Article 230-71(b) says that you have to arrange all the disconnection means in a way that you can switch them off with no more than six operations of the hand. (Why is six such a recurring and magical number in the NEC?) Since there is no reference that points to the six-motions-article, how do you find out about it? The answer is: Inspection time.

Inspection time is trembling time unless you know the Inspector well and he knows you well. (Looking at some "normal" AC installations in my area, I wonder, at times, exactly how well did they know each other?)

Strictly following NEC Article 690 when installing a solar system is, simply said, impossible. Either you install so many switches, fuses, and other means of disconnection along the line that there is no voltage coming out the other end or you install grounding conductors as thick as a rattlesnake, costing you more than the rest of the equipment altogether. And, even if you did your very best, here comes the Inspector.

I had never met him before. It was Monday morning. Even though it was a clear and sunny day, my mood was cloudy. It was a big house, a big system, and a big chance I took because the new 1996 NEC was just out. Not that it scared me more than the last Code, but a new inspector and a new code. . . .

It was to be an "a.m. inspection" and the Inspector showed up at 11:30 a. m. (Needless to say, I had been waiting since 8:30 a.m.) After a good handshake, he went to work. The AC part went very well. He plugged his little tester into every receptacle, threw every switch, but stopped dead when he came upon the DC receptacles. His little plug-in device would not fit. "That's the purpose," I said. "AC devices should not fit into DC devices." One could tell he didn't like it a bit. I lent him my tester and grudgingly he had to accept the fact that the DC receptacles also worked.

Now off to the power chambers. He wanted to know the meaning of life. Patiently, I gave him the grand tour, wondering how much he really knew. After I finished, he was silent for a while. Then he started counting: One, two, three, four, five, six, seven. He started again and I started to wonder. Yes, he came up with seven again. Well it didn't take a qualified electrician—let alone an Inspector—to tell that there were seven means of disconnection. (I had explained every single one of them.)

"You can't have that," he said.

"Have what, Sir?" I asked.

"Seven," he said.

"Seven?" I asked.

"Yes, seven!" he said.

Well, there it was. Seven! Of course, I argued that nowhere in Article 690 of the Code did it say that I could not have seven means

of disconnection. But then I got the finger. I was informed that this was basic knowledge. Six! Six is all I could have. Everybody knows that. And with that he left and I got "red-tagged."

How could those six motions of the hand have escaped my otherwise superb knowledge of the Code?

To Code or not to Code is often the question when it comes to photovoltaic installations. The Code only wants to protect but does not seem to be inter-

112

ested in serving. And while it is important to prevent you from starting a meltdown in your battery room or creating a hydrogen explosion, it often addresses situations which are simply impractical or outright damaging to the efficiency of your system.

Certain parts of the Code are recommendations, certain parts are THE LAW, certain parts deal with requirements for which no equipment is available on the market or, unless you buy 2,500 feet of a certain type of wire from a certain manufacturer, are impractical or too expensive.

Not only do you have to deal with certain types of cables which are hard to get but most simple switching devices, fuses, and breakers are not rated for DC use. Fortunately, there are some available, but in many other respects they might not suffice. Switches, fuses, breakers, and connectors usually have a temperature rating. This means that they are designed to operate up to a maximum temperature. Changes in temperature can be caused by outside factors (like the heat on a metal roof) as well as inside factors (heating up with high currents). By now we know that DC currents are usually higher than AC currents. (Remember the formula: **Volts x Amps = Watts.**) This means that DC devices have to be able to withstand higher temperatures than AC devices. For this and other important reasons (DC switches spark stronger than AC switches, etc.) most AC and even some DC-rated devices cannot be used for certain applications. The proper switches and fuses that conform with all these requirements are often hard to get and rather expensive. In the end, it is the Inspector's call as to how far he wants to pursue some of the hard-to-comply-with sections of the NEC.

If you look at the Code requirements for a

conventional home which is usually based on power requirements of up to 200 amps at 220 volts (which is about 44,000 watts), one may ask why should a solar home be wired according to these requirements? A legitimate question, especially when it comes to a small mountain cabin or the like.

My recommendation is that, if at all feasible, build to Code. That may mean that you have to supply the kitchen (however small it is) with two separate circuits for small appliances, put a dedicated circuit in for the washer, for the bathroom power, etc. Of course, if the small cabin only needs six outlets and three lights, what can the Inspector say?

In the end, it comes down to what your Inspector deems to be acceptable. Some Inspectors do not want to deal with photovoltaic at all and have very little to say about it.

Some Inspectors do not want to deal with photovoltaic, but have a great deal to say about it. Very few Inspectors really accept photovoltaic as a real alternative solution to grid power and thus invest time to train themselves properly so they can deal with the situation adequately.

The best way to avoid unpleasant surprises is

to get in contact with the Inspector for your area ahead of time. It is wise not to challenge him or his wisdom (he may just say: "Follow the Code."), but to give him your idea of what you want to do and ask him if that would work. It often does. Then you always can remind him that it was "his suggestion" that you followed.

Whether you plan on building your own solar home or having it contracted out may change after you read this chapter. The do-it-yourself times are certainly over as far as electric is concerned. However, if you happen to run into a solar installer who also is an electrical contractor and who is willing to work with you, there is a lot you still can do yourself. I highly recommend that you be part of all phases of the construction of your solar home, particularly when it comes to installing the solar system, because this way you will know your system and its components well, which will be of tremendous help when troubleshooting time comes. Of course, we will talk about this later.

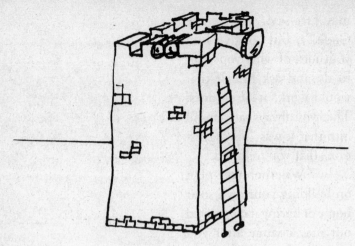

Chapter Nine
MONITORING
THE SYSTEM
Watching and Guarding

Now let's assume that everything went well. Which is a reasonable assumption, right? Every component is installed and, after a few setbacks and adjustments, they seem to work. The solar installer has packed all his tools into his truck and has disappeared over the horizon. The sun is setting and silence is descending with the veil of darkness upon the lonely desert.

Happy and content, you turn around and walk back into your newly equipped house and, in anticipation of a quiet evening watching your favorite TV show which you have been unable to do for many weeks, you flip the light switch on. (Of course, I know that I should not assume that the readers of this book watch every soap opera and hence, my story should reflect a more healthy and alternative lifestyle. But I have already made so many adjustments to be politically correct that by not being politically correct in a funny and—sorry—entertaining way, I think I am even more politically correct than I would want to be. Got it?)

Silence and darkness is all you experience outside as well as inside the house. The AC lights do not come on. You walk into the utility room and notice that it would have been smart to install at least one DC light inside this room, wired directly off the batteries.

You search for the flashlight, remembering the warning not to use any open light near the batteries and look at your array of switches, fuses, and meters.

What went wrong?

I will leave you in the dark about what went wrong for now but I promise you will hear the rest of the story in a later chapter which will be dedicated to troubleshooting alone. (**Chapter Twelve**, p. 171.) The point here is that the art of monitoring can tell you in a few seconds what went wrong.

Art means mastering your trade and, at this particular point in time, you are most likely an apprentice. So what are you looking for when you monitor your system? You are looking for indications that your system performs to its specifications, which means that it delivers what was promised to you by the manufacturer, dealer, and installer.

You are also monitoring your ability to live with it, learn its cycles and its headaches, and know how to avoid them or plan around them. Monitoring should be a scheduled and regular undertaking. A quick look in the morning and one before you go to bed will usually do, if you know what to expect.

At first you try to understand the voltage. You will find that early in the morning volts are low, in the early evening they are high, and in the early afternoon they are highest. This is the cycle of recharging.

On a clear day you may observe the voltage reaching 14 or more on a 12 volt system and 28 volts or more on a 24 volt system. This indicates that the batteries are reaching their "gassing" level and, if they are new and functioning well, their fully charged stage.

If you come home on a day like this, let's say one or two hours after sunset, after the charge controller has stopped charging, you will be able to read the true voltage of your batteries. At this time, the charging process inside the batteries has settled and the voltage has established itself at about the level the batteries can hold for some time.

When the battery charge reaches above 14 volts (the *gassing point* or *gassing voltage*, usually 14.2 to 14.6 volts), the charge controller will stop the process and turn itself off for a time. You will notice that the voltage starts dropping immediately a few decimal points below 14 volts, at which time the charger will come on and try again. This process, called *trickle charging*, will continue until sunset, if no electricity is used, or until the voltage drops drastically because some user is turned on.

As we learned earlier, the batteries need a higher voltage than their nominal voltage to fill up. But this does not mean that they will keep or hold the voltage they are filled with. One hour to two hours after sunset, their voltage level will sink from 14 plus volts to about 12.8 to 13 volts. If this is the case when you come home, you know you have full batteries. If you do not see this voltage but a lesser one, you know that:

118

a. It was a cloudy day.

b. You left the lights on in the pantry.

c. Your kids watched TV in the afternoon.

d. Last night you used too much power (battery abuse).

e. If none of the above, you have a problem.

Now your problem may be that you bought the system with the house and you were told that everything was in excellent shape. This might be true as far as everything else is concerned, but your batteries are probably as old as the house.

If, during the daytime, the voltage reaches 14 volts or more in no time, but at night sinks below the floor as soon as you turn the entry light on, your batteries are most likely on their way out. The voltage you monitor during daytime is called a *float charge*. This voltage is trying to charge your batteries but the batteries are too old and weak to absorb it. All

day long the surface of the batteries are "cooking" with high voltage and the bottom is low and empty. A battery acid tester, the same one you use on car batteries, will tell you after a long and sunny

day that your batteries need to be recharged.

This situation on change of ownership of solar houses seems to be the case more often than not. ***Beware:*** Do an acid test before you buy and get written assurances about the age and condition of the batteries.

Late at night before

you go to bed, note your system voltage and check it again first thing in the morning. In the morning, it should be the same or even a little higher than it was at night. The reason for this is that, while you use electricity from your batteries, they tend to drop momentarily to a low voltage and, after use, "recover" to the higher voltage minus what you used. If you did not use anything during the night, they should have recovered to a slightly higher level than they were at before you went to bed.

Sophisticated monitoring gadgets will show you several different voltages. They will show the array voltage, the battery voltage, and the load voltage. You will note that the latter two are more or less the same, except in winter when the temperature of the batteries may drop with the outside temperature. If the battery temperature reaches the freezing point, it may lose up to 50% of its capacity. This will reflect itself in a lower voltage. The same is true, even though not as drastic, for very hot summer days. Batteries like a temperate climate—50 to 75 degrees Fahrenheit suits them just fine.

The first voltage mentioned, the array voltage, might be as high as 16 to 20 volts on 12 volt systems and 38 to 40 volts on a 24 volt system. This is called *open circuit voltage*. This is the voltage your panels show with nothing connected to them. Usually this is of no great interest to you. However, it can show you whether there is a short in the array circuitry.

Watching the voltage should be a regular exercise. You can discover trends of low battery status that may shut off your inverter and with it all the exciting, even if not politically correct, things you can do with AC power. DC equipment may also suffer with low voltage. Usually answering machines,

DC TVs, and other electronic equipment are quite tolerant as far as the voltage is concerned. But at some point (and who is to know when), they also cave in and often suffer some damage. DC lights don't care at all, they just burn a little dimmer.

The next thing to monitor are the amps. Again you may monitor the charging amps, the amps to the inverter, and the amps used in the DC circuits—the DC load. The most important amps are the charging amps. You will see that they are very low in the morning, peak at noon, and decrease until sunset. They are a direct reflection of the sun hitting the panels. They can tell you whether the day is as clear as yesterday, provided you check exactly at the same time; they can tell you whether your panels need seasonal adjustment, if your peak amps measured at the same time of the day decrease over a period of weeks or months. And they can tell you whether your panels perform to their specifications or not, provided they are adjusted for the right seasonal angle.

Let's say your panels give you a peak reading of 15 amps at one o'clock daylight savings time. (Remember that noon is at one o'clock in summer because the time is adjusted one hour ahead during daylight savings time. Spring ahead, fall back.) If your array voltage peaks at an earlier or later time than noon (one p.m. or 13:00 in summer) it shows that your panels are sitting at the wrong angle. The correct angle would, of course, change every day because the sun rises higher above the ho-

rizon every day after December 24th and sinks lower every day after June 24th. As a compromise, you should adjust your array twice a year to the angle of your latitude minus 15 degrees in summer and plus 15 degrees in winter.

To find out whether the panels perform to specifications, you multiply your amp reading by your voltage. You should get the watts your array is charging with. (**Amps x Volts = Watts**) On the back of newer panels, you will find their nominal voltage, their watts, and their amps. Let's say the label reads 15 volts. Multiply 15 volts with 15 amps (the amp reading at noon) and you will get 225 watts. Now divide 225 watts by the number of your panels and you should get the watts for each panel. If this gets you close to the labeled watts, you are in good shape. The reason you use the voltage on the label and not the one on your voltmeter is that you will most likely

have lower voltage on your voltmeter due to your battery status (plus, the manufacturer used this voltage for the labeled watts).

If your amps are labeled on your panel, all you have to do is divide your reading by the number of your panels and check the result. You should be close to the label, but you most likely will never reach the exact same reading because of all the connections, cable lengths, breakers, fuses, and switches that are installed between your array and your meter. Also, if the batteries are near their fully charged stage, the "back pressure" from the fully charged batteries will decrease the incoming charge power and the amps will read lower at this time. So it may take several attempts to get a proper reading on the actual performance of your array.

The next things to monitor are the DC amps or load amps. Remember, DC amps are five or ten times higher than what you may expect of any AC load. Turn on your 15 watt AC compact fluorescent light bulb and you will see your DC-amps indicate up to 1.5 amps. Turn on your AC TV set and you may see as much as 7.5 amps. Now, turn on your hair dryer and observe as much as 120 amps on your meter. If you were doing light carpentry for most of the day, using a circular saw or a small table saw, a power drill, and a sander, you may be surprised to see an unexpectedly low voltage by the end of the day. But if you were watching your amp-meter, it would become clear to you that you used a lot of amps running these machines.

Again, it is not wise to use your solar power for extensive construction even though your contractor may try to persuade you otherwise. Observing the amps going up and down while you turn on all your saws and drills may give you an idea of what you are using.

Monitoring amps can be very educational, but at times it can also be confusing. If you own an inverter which indicates AC amps in use, you may get confused because you may have forgotten that AC amps are different from DC amps (remember, in a 12 volt system you need about 10 amps out of your battery bank for every one amp the inverter uses), but once you figure it out, you will know quite well why your system is low at certain times and not fall for some of the explanations I often hear: "No, I didn't use much yesterday, I only ran a few washing machine loads and ironed for half an hour. The iron only uses 10 amps!"

A meter that gives

you an amp-hr reading can tell you exactly how many amps you used during that period. If you know what your battery capacity is and you know how much you used, you will know how much you've got left in them.

Let's say you have four 6 volt batteries at 220 amp-hrs on a 12 volt system. You know that you have two pairs of 12 volt batteries which will give you 440 amp-hrs total capacity. (Two 6 volt batteries at 220 amp-hrs make one 12 volt unit at 220 amp-hrs.) You do not want to discharge your batteries below 80% of their capacity. So you have about 350 amp-hrs available (80% of 440 amp-hrs). After you finished building the rocking horse for your three-year-old, you noticed that you used 150 amp-hrs. Since you only

worked after hours, which means after sunset, you know that you have only 200 amp-hrs left in your battery bank. That is probably enough to get you safely through the Super Bowl but, if you planned on surfing the Internet for the rest of the evening and printing out some exciting articles while your wife surfs the network channels on TV, you may have to consider pulling straws to see who gets to use an electrical appliance and who will have to read a book instead.

Of course, you also have to consider your recharging power and a possible change in the weather, with clouds moving in. So leave some reserve or you may have to run the noisy generator which also pollutes the air.

Amp-hr meters are a little tricky to use. First you have to tell them what capacity your battery bank has, so they know what to base their calculations on. They usually reset themselves when they reach battery peak voltage, which is around 14.2 volts in a 12 volt system. Now they start counting down again if you use some energy. It gets confusing if you didn't tell them what the maximum amp-hrs of your batteries are, and they keep on counting up after they reach the setting point. Now you end up with a figure that may not reflect your actual battery capacity.

There are many other things you can monitor, and you can become a scientist with complex statistics. You could even have your computer draw out a graph on every aspect of your electrical life, if you have one of those expensive monitoring devices. But most important is to find out the rhythm of your system and how it relates to <u>your</u> rhythm. And don't get paranoid when the voltage starts dropping while the TV or the washing machine is running. It may not be time to pull the old wash board out yet. Remember that what counts is to which value the voltage recovers after you stop using the appliance.

I sure hope that you are not approaching the valley of your learning curve or getting idle on some flat learning plateau because this is the stuff your daily solar life will be made of. You need to know at any given time what condition your system is in. And all it takes is a quick look at your metering devices to make a judgment call. "Hey kids, no TV tonight, pull out the books and start reading again, and, by the way, have you done your homework?"

So what did we learn

in this chapter? We learned that it is a wise thing to invite your solar contractor in for a cup of tea after the work is done. (Don't serve coffee. You know these guys always like to show off their superior knowledge and never stop talking.) It may take him only a few minutes to fix what may cause you a night of despair and a long distance call the next day.

We also learned that the more you know about your system, the fewer surprises you will have when it's time for the Super Bowl.

We learned that knowing why the voltage is low in the morning, high in the evening, and highest in the afternoon can tell us something about our family, the weather, and other potential problems. We learned that a *float charge* has nothing to do with a cover charge and a *trickle charge* is like trying to top off a champagne glass. And just as a good glass of beer has to sit for a moment to let the foam settle (at least that's how it is done in Germany), so do our batteries have to sit after sunset in order to give us the exact state of their charge.

We learned that watching the charging amps from the solar panels can tell us whether we are getting what we paid for, and watching the amps while using a hair dryer can tell us that a clean and dry head sometimes needs an awful lot of energy. But, most importantly, we learned that monitoring our system will make us experts on our solar power plant and that, once we master the skill, we may not always be able to get from it what we want but at least we will know why.

Chapter Ten
TIPS AND NAMES
There is Always
a Back Door

This chapter will cover tips on how to get what you want without having to double the size of your system. For almost every situation that may call for an expensive solution, there is an alternative way it can be done. You also may have no-

ticed that I have not yet named names. Well, this will change now—it's "payback" time. You will hear about products and companies and my experience in the field and on the phone with them.

Actually, I regret that I didn't write this book in the

plural. It would have had much more weight and profundity to start a sentence off with: "Well, we found that...or our experience is that . . ." But I may still be able to do that because I am not alone. The experiences of other solar contractors and suppliers in the area will be reflected in this chapter.

Some practical hints and tips may seem familiar to you. That is because I mentioned them earlier. (Hey, that was cheap!) Other things are just common sense and may also seem familiar to you because they just popped up from the collective knowledge. But sometimes it is good to have those things compiled and readily available even if they seem banal to you. We will go through your future home and talk about things in geographical order. Let's start with the kitchen.

KITCHEN
Toaster
You think you need

one? But you can't run it on

your system. The old prehistoric metal plate with a few holes in it and a handle on it that you heat over the stove will do the trick. The plate is actually a double plate with spacers in between to distribute the heat better. You can still find them in hardware or kitchen supply stores. But watch the heat and regulate the gas flame down because the heat accumulating under the plate combined with bread crumbs falling through the holes tends to discolor the enamel on your stove top.

Well, that was easy, and who would have thought that the kids can have their peanut butter and jelly toast in the new solar home? Of course, toast with tahini and a knife tip of white miso also tastes excellent. (But I won't start a sermon on dietary habits right here.)

Stove Exhaust Fan

Let's stay in the kitchen because it is really the women who design the house. At least that is OUR experience! The more contemporary house designs often have the kitchen combined with the dining area and are open to the living area. This will make the smell of Sauerkraut and Bratwurst move into your expensive oriental rug on the living room floor. You need a means to exhaust those smells. Also the toast left on too high, creating those dark clouds and indoor pollution, will trigger the smoke detector in the open hallway to the bedrooms.

A range exhaust fan usually does not use too much power, except of course that it is most likely an AC fan/light combination and the incandescent light bulb in this hood may use a fair amount of energy. But you can avoid it altogether by installing a small window over the cooking range. This not only inspires you while cooking, if the views are right, but the cracked window really draws out the cooking smells. (You do not have to crack the glass though.)

Oven/Range with Electronics

Electric ovens and ranges use up to 1,500 watts per burner and are not recommended for photovoltaic systems. The alternative is using propane gas ranges. Most modern gas ovens and ranges are designed using electronic circuitry to control thermostats and ignition. These circuits may cause a

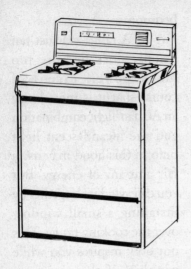

wired off one of the kitchen's small appliance circuits. All you need to do when using the stove is flip the switch. Of course, you will not be able to use the built-in electric clock on a continuous bases, because it will also stop when you turn the power off. To most people, however, this seems to be a minor sacrifice.

But here is one thing to watch out for: several of the newer ovens use a heat coil to start the oven (instead of the old-fashioned pilot light) which, like everything that produces heat, is a high consumer. Ask your appliance dealer about this before you have it delivered and installed.

constant phantom load and keep the inverter working overtime. The simple solution is: No electronics. Either you get an older stove that does not need electronics or you buy a new one which does not have any electronics in it. One brand is called **Peerless Gas Range** and is available through **Alternative Energy,** 1-800-777-6609, or a local appliance store.

But if you have a special design, color, and brand in mind, you can install a switched receptacle for the stove. The receptacle can be

Fridge/Freezer

No AC fridge is an option for a solar system. Modern propane fridges and freezers perform very well and are reliable. Their price, however, is still something to complain about. They cost about twice as much as their

AC companions. The same is true for DC fridges. DC fridges, in particular those from **Sunfrost**, 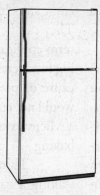 are built to save energy. They are heavily insulated and run with special compressors that have a low surge and use very little energy. But they still use enough that you have to consider installing extra panels and batteries to compensate for their use which makes them even more expensive. However, they perform extremely well, are reliable, ice up very little, and keep temperatures constant.

If you opt for a cheaper solution, there are several fridges on the market which are used in RVs. They often come in three-way arrangements, 12 volt DC, 110 volt AC, and propane. Usually they are smaller and fit under counters. Most commonly they have to be installed inside a cabinet and be vented to the outside as well as get their combustion air from the outside. Typical brands are **Norcold** and **Dometic**.

Of course, there is another solution to keeping your food cooled: the old-fashioned pantry. If built on the outside of the north side of your house and properly insulated and equipped with a low vent hole on one side and a high vent hole on the other side, you will have a cool place for most food items all year round. This lets you get away with a smaller fridge for those items that need it really cold. You may be able to turn your fridge off altogether in winter and use the pantry as a walk-in fridge.

You may have to give up some of those unhealthy habits, like drinking your water at temperatures that would slow down almost any subatomic movement (to say nothing about your stomach lining). But that may be a small price to pay.

Small Kitchen Appliances

Provided that you have a properly sized inverter, most kitchen machines have no trouble operating on a solar system. Even big dough mixers and grain grinders run without any trouble.

The infamous microwave, however, may need

a lot more juice than you would expect. Most use around 1,500 watts. Besides, there is the discussion about their safety—in which I will not get involved here. (They're supposed to destroy your cell membranes and will eventually make the whole human race mutate to be-

come amoeba-like creatures.) They can be used for short-term applications, like heating food, etc. But strictly because of power concerns, I would not recommend using a Microwave for cooking or baking.

Thinking about installing a garbage disposal? What's wrong with a nice compost pile? A garbage disposal needs a dedicated circuit because it is a permanently installed motor and a quite power-hungry one. Of course, you can run it for a short while, but it will weigh heavily on your load calculation.

Dishwashers

I recently read an article interviewing famous Chefs. About half of them mentioned that one reason they became Chefs was that they hate to do the dishes. I have to second that because I also became the cook in the house for the same reason.

A dishwasher uses quite a bit of energy. Not as

much as a clothes washer but, still, you need to consider the inconvenience of doing the dishes by hand vs. the extra energy you need to have a machine do the work. On the other hand, dishwashers may save you water and, if you do not use the drying phase, they might not consume too much energy.

Alternative considerations do not need to be discussed here. (I don't want to get in trouble with husbands and kids.)

Kitchen Lighting

Because of the special lighting needs in the kitchen, you often find under-cabinet lighting. Most of these lights are fluorescents with a very low wattage. Starting those lights on your inverter may become a problem. Most inverters need a minimum wattage to start up from the stand-by mode to the inverting mode. Those lights usually use so little energy that the inverter will not stay on to run them. You either need to switch several lights on at once or to use under-cabinet light fixtures with halogen bulbs in them. Halogen bulbs usually use more power than fluorescent lights and have no problem starting up the inverter.

BATHROOM
Hair Dryer

Hair dryers use between 1,000 to 1,500 watts, which is, by solar standards, a prohibitive amount of energy. Running a 1,500 watt hair dryer for five minutes consumes about as much as your 25 inch TV uses in one and a half hours.

The need to have well-dried and well-styled

thing in common: they mostly use rechargeable batteries. These, of course, create a problem because they need to be plugged in and recharged all the time—or do they? No, they don't. I have not tested electric shavers yet, I let my beard grow. But electric toothbrushes, once charged, last for several days until they have to be recharged. The ultrasonic tooth-

brush from **Braun** lasts up to a week on one charge when used by two people. Electric shavers probably have a shorter lifespan, but they also should last several days before

hair will not be discussed here except to say that these things seem to have become more important in recent years. The old dread-lock fashion is slowly fading, although it lasted longer than bell bottoms.

An electric "hair brush" made by **Vidal Sassoon** uses "only" 150 watts. It may take a little longer to dry your hair with this appliance, but you can also style your hair at the same time.

Electric Toothbrushes and Shavers

I don't mean to imply that they are interchangeable, but they do have one

you need to recharge them.

Bathroom Lights

When it comes to lights around the bathroom mirror, you will have to forget about those fancy backstage-make-up light fixtures using half a dozen light bulbs. Imagine, six times 75 watts! That's a little bit too much for your system. But any fluorescent fixture or compact fluorescent light will do. (See the section on *General Lighting* on page 150 for a discussion of problems with fluorescent lights.)

GFCIs

GFCI stands for *Ground Fault Circuit Interrupter.*

These are receptacles usually with one red and one black button which you find installed in most bathrooms and above kitchen counters. They are supposed to prevent you from drying your hair while sitting in the bath tub. The moment they sense a possibility that any electricity could reach the ground (i.e., by means of your body), they disconnect the power. Certain brands do not agree with certain inverters and will not work. If you encounter this, do not despair. Just exchange them with another brand.

LIVING ROOM
TV, VCR, Stereo

The living room with its whole array of entertainment gadgets, in the context of the newly learned term "phantom load," seems to be a nightmare. But, it does not have to be. Certainly, any of these electronic devices will draw a phantom load when left plugged in, even if turned off. Now you picture yourself on your knees unplugging the TV while plugging in the stereo, then unplugging the stereo and plugging in the computer, and finally unplugging

135

the computer and plugging in the VCR and the TV. You are getting the picture, right?

Here is the solution for this special workout program. It is called a *power center*, or *computer control center*. This is a device about half the size of a VCR with several plug-in receptacles at the back and an individual control switch for each receptacle plus a master switch for all the receptacles. This power center costs between $14 and $20. It usually comes with some surge protection built in. If you want to watch TV, all you have to

do is turn on the master switch and then the individual switch labeled TV. All other devices remain disconnected until you switch them on at their designated switch. At night, when every thing is said and done you just throw the master switch and everything is off at once.

Now if it is that easy, there has to be a disadvantage. There is. The only gadgets that will create potential problems are those which need a constant feed of "stand-by power" to keep certain memories activated, such as certain TVs which go through an initial set-up mode to memorize active channels and, of course, the VCR.

Statistics have shown that the average American is not capable of programming a VCR properly (that includes several Presidents in the White House). Most VCRs will not retain their memory once they get unplugged. Therefore, you will have to reset the clock and, if applicable, the date and time zone, and answer all these questions concerning your private life over and over again. Considering this, you may decide to give up on the solar concept altogether, especially if you own a newer VCR Plus. (I have not man-

aged to get my VCR Plus working properly. If I set it to record my favorite show, I always end up with one of those movies, like **Zombies At The Supermarket** or **The Gravedigger's Nightmare,** which are broadcast at three in the morning.) But if you are not President of this country and have just a bit of faith left, you should be able to overcome this par-

ticular obstacle in one way or another.

As mentioned, these devices do not use as much as an electric iron but still will consume quite some amp-hrs because they run a little longer than it takes to iron your shirt for Sunday church. TVs you find usually advertised as 60 watts. However, I have found that an 18 inch TV uses more than that and a 25 inch TV uses about 90 to 100 watts. Add your VCR with another 30 watts and you are up there. Watching a two hour movie will consume about 250 watt-hrs. There is nothing you can do except watch less or buy a smaller TV.

With stereos the situation is different. A big stereo uses more than a small one. Big speakers can bring you to the edge of your nerves and of your power reserve. But if you accept a little lesser quality, you can get away with considerably less using a portable device. The advantage is that, besides having relatively good sound quality, they run on DC. And if you find one that runs on 12 volts, you can connect it directly and bypass the inverter. If you have a 24 volt system, all is not lost, because a converter can be used to let you have your

music from the DC source.

The problem with humming and buzzing depends on the brand of stereo and your musical ear. If you are plugged into AC and you are using a modified sine wave inverter, you may get a buzz. Some brands buzz less than others. I found that Sony has the least buzz, audible only on low volume settings. If you connect directly to DC, however, you will not get any buzz at all

on any brand because you bypass the inverter, another advantage of a small stereo system.

Computers

If you have a computer, it might be too late for this suggestion, but if you plan on getting one or on upgrading to a new one, consider this.

Most computers work just fine but some don't. I have heard of several occurrences where an **Apple** computer's power supply could not handle the modified sine wave and burned out. I checked with **Apple** as well as with **Trace Engineering** (the inverter manufacturer involved) and both had no idea what I was talking about. Computer repair places, however, know that **Apple's** power supplies are "rather simple" and could see the possibility of them failing in other than conventional power situations. I know of one case where the whole "mother board" of an

tioner, which cleans up any "dirt" and rusty edges within your electrical current. These gadgets cost about $1 per watt and you may consider them in case you want to use a Mac computer.

If you have a real computer, and not just one for video games, you might want to print from time to time. The only printer that could create problems is the Laser printer. Laser printers use heat in their printing process, and you know what heat can do to your batteries. Inkjet printers deliver a high quality and use much less energy.

Apple desktop burned out after only three months of use. Needless to say, the manufacturer did not honor any warranty.

Another consideration is the energy use of a desktop computer vs. a laptop computer. A desktop uses around 90 to 120 watts, plus whatever monitor and printer you are using. You may end up with up to 200 watts, while a laptop operates with only 40 to 50 watts. **Apple** laptops do not seem to have any problem running their power supply on a modified sine wave. Of course, there is a device which is called a *line condi-*

Fax

A fax machine can be a real problem if you need to have it on at all times. It uses a high amount of "stand-by" power. I use a fax machine the expensive way (expensive for the caller!).

Whenever somebody wants to send me a fax, they have to call me and I turn on the machine. (This is still pretty common practice in Europe, but it has made for some unhappy callers.) Since the explosion of e-mail, however, most of these problems have been solved, because you can get e-mail any time. It gets stored at a receiving facility (computer) and you call in from your computer and download your e-mail whenever you want.

Answering Machines and Cordless Phones

Answering machines naturally have to be on most the time. Therefore, to avoid the "pain-in-the-neck-load" (*phantom load*), you will have to find one that can be connected directly to DC. Fortunately, there are several brands available that come in 12 volt DC. **Phone Mate, AT&T,** and **Panasonic** are some of the manufacturers that sell models which run on 12 volt DC. (They and other manufacturers also sell models that run on 12-14 volt AC, which would not work for our purposes.) The way to connect these is to get a new power cord at any Radio Shack or other electronics store and replace the cord it came with, which has the (usually) black transformer cube that plugs into an AC receptacle. Watch out that you do not cross polarity, which will blow up your machine and void the warranty. (Consult your electrician if you are not sure.) If you cannot unplug the cord that came with your device

140

because it is hard wired into the device, you will have to cut the cord just short of the transformer (the black cube) and, after you carefully check polarity, connect it to your DC source. Needless to say, this would be a nice alternative to an answering machine.

A quick word on wiring techniques here (this is supposed to be the next book). Make sure that you or your electrician run those phone lines and speaker wires as well as TV lines at least 12 inches away from any AC line. This can create quite a challenge, as you will find out. If you cannot avoid an AC line, make sure it is crossed at a 90° angle.

this voids the warranty, but I have done it and gotten away with it because the answering machine had no problems for the year it was guaranteed (which today seems to be the exception rather than the rule).

If your phone company can provide you with an answering service, usually for under $10 per month,

Fans

There are several DC fans available which you can find in solar supply catalogues, such as **Real Goods, Alternative Energy, Jade Mountain,** etc. There were a few overhead fans available but they seem to have disappeared. AC fans, especially oscillating and overhead fans, use far more energy but, on hot summer days, you also may have far more energy available. A good four

blade fan uses between 0.5 to 1 amp and 110 volts, which amounts to about 50 to 100 watts, which is acceptable unless you plan to run them all day.

Fan and Light Controls

This is a somewhat tricky issue. A fan control as well as a light control, commonly called a dimmer, puts out resistance to eat up some of the electricity which, in effect, slows down the fan or dims the light. I use the term "eating up" because it is true—the electricity is gone, used, and transferred into heat. So even with a slower fan and a dimmer light you still use the same amount of electricity you would use with full speed and bright lights. This is just something to remember when you use these controllers.

However, if you use compact fluorescent lights, in most cases you cannot use a dimmer at all. Remember, fluorescent lights don't "burn," which means they do not create light by super-heating tiny wires inside a bulb so they start glowing or burning, which is what incandescent bulbs do. Only the absence of air inside the bulb prevents them from burning out. If you use a dimmer, they either burn hotter (brighter) or less hot (less bright). Fluorescent bulbs create "cold" light by letting electrical current move molecules of helium, neon, crypton, or similar gasses. They only create light when a certain current is flowing through them. They are either on or off, there is no in-between. That's why you can't dim fluorescent lights. However, **Phillips** just developed a compact fluorescent bulb that can be dimmed. Its called **Earth-light Dimmable**. Dimmers are also available in DC currents. However, they are considerably more expensive.

BEDROOMS
Lighting

There is not an awful lot to say about your bedroom. (Nice choice of furniture.) Bedside table lamps can be halogen, except that some people object to that,

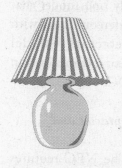

claiming that they have a strong electromagnetic radiation which should not be close to your head.

Electromagnetic fields, which those halogen lights radiate in higher quantities than other lights, seem to create problems for some sensitive people who report headaches and other forms of discomfort. I have had customers who requested that there be no power lines inside the walls behind their bed. But if you make sure that you do not have any phantom loads to make your inverter do overtime at night, you should not have to worry about these fields. When the inverter is on stand-by, there will be no current going through your walls.

When you install closet lights, put the switch inside the closet so that you know that you turned it off before closing the door.

Smoke Detectors

In most states, the fire protection code requires that there has to be one smoke detector inside every bedroom and one on the outside in the hallway between bedrooms. This amounts to quite a number of smoke detectors in the house. They all have to be interconnected (a *multi-station system*) so that they all go off if one gets triggered by

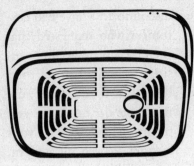

have a higher voltage than 12 DC. Connecting them in series seems to mean that the first one burns out occasionally, so it would be better to convert to the lower voltage first.

They are available from **System Sensor**, 3825 Ohio Avenue, St. Charles, IL 60174, 1-800-736-7672, Fax: 630-377-6495. They carry only one model that can be interconnected with other detectors (Model 2012B) and it comes only in 12 volt DC.

some smoke in the house. They may have to run off the house voltage (110 volts), and you know what that means—yes, the inverter will have to be on all the time to feed those little sirens. There usage is so low, however, that they would not activate an inverter in stand-by mode and you would have to keep your inverter in a constant "On" position.

There are some DC smoke detectors on the market which cost considerably more than ordinary 110 volt devices, but they would eliminate the problem of running the inverter all the time. As a *multi-station system*, they are only available in 12 volts, which is a consideration if you choose to

DC Receptacles and Switches

The NEC requires that DC receptacles be different from AC receptacles so that a mix-up is not possible. They also have to be UL listed. (UL listed means that they have been tested by Underwriter Laboratories and found adequate for the supposed use.)

There are cigarette lighter-type receptacles available. Many DC devices come with plugs that would fit those receptacles. Some people hate them because they make funky connections and some people like the convenience of not having to change plugs all the time. However, these receptacles are not UL listed and would most likely not pass inspection.

The most common solution is to use receptacles that are not commonly used in residential applications but are UL listed. A typical example is a 15 amp, 240 volt rated receptacle that is also referred to as the "Chinese Eye" because it has two horizontal slots where the normal AC outlet has vertical slots. They come in the same size and colors as the normal AC models.

Switches have to be DC rated because DC creates a bigger spark or arc when switched. Although I have used AC rated switches over the years and had only two go bad in six years, I have to use DC rated switches now because the Code requires it. Considering that they cost $4 to $5 per switch vs. $0.5 to $1 for the AC models, there is quite a price difference. But usually there are only a few DC switches in the house.

UTILITY ROOM
Washer and Dryer
Buying a washer for

a solar home can be an expensive choice. If you want a real water and electricity saving model that has also excellent washing qualities, you have to spend around $1,000 or more. The top model is the **Staber System 2000 Washer**. It has a very low starting surge which lets it run even on small inverters, using only 4 to 5 amps.

Another model is the **White Westinghouse,** a front loader that also uses only about 5 amps but has a higher starting surge and uses more water than the **Staber**. You will only find those second hand. The **Maytag Conventional** is a stacked washer and dryer that also works well on small inverters.

Considering that American washers have not evolved since the 1950s and still use up to 70 gallons of water and a lot of power to distribute the dirt more evenly within your clothes, the above models are revolutionary. By European standards (after which they are modeled), they are the middle of the line. But the industry is slowly starting to recognize their handicap and newer and better models are in the works everywhere. So you may want to inquire about newer models at your appliance stores. Unfortunately, most salespersons also seem to be leftovers from the early 1950s and do not understand what your inquiry is all about. So please be patient and don't yell at the poor underpaid people (the way I did). They are only doing what they (barely) get paid for.

We have found one problem with most of the newer washers. Their electronic controls do not run on modified sine wave inverters. The models mentioned above are either designed for these inverters (**Staber**) or happen to have no problem with this form of electricity. So don't throw away the box of your new washer; you might need it again.

Dryers can only be used when they are operated by gas, which means the heat is created by gas and the tumbling drum is operated by an electric motor. The interesting thing is that, even though the dryer runs on gas, it still uses more electricity than the washer, which is about 5 to 6 amps or 600 watts.

Considering that it often takes longer to dry clothes than to wash them, you know why in a minute I will suggest an alternative means of clothes drying. There is this device, it is made from nylon, it weighs about a pound, it costs about $2 to $3. It comes folded and needs to be assembled. All you need is two to three trees or posts and, *viola*, you've got your clothes dryer. It is also known as a clothesline. Of course, towels get rather stiff drying on this line. But if you rub them and shake them after they dry, you will be surprised. (Now I will not take credit for this discovery. The credit goes, with the full weight of the work involved, to my wife.)

DC Pumps

Well, whether you have a well or you don't, which means you catch the rain when it falls and store it inside what is known as a cistern (no, not the tiny ones you flush your toilets with), you need pumps to either haul the water from the depths of the earth or to create pressure in your water system or both.

If you have a shallow well, down to about 200 feet, a **Solartek** pump or a **Solarjack** pump will perform well for a reasonable amount of money ($600 to $700). For deeper wells, you need more powerful and more expensive pumps like the **Solarjack SCS** pump that pumps as deep as 800 feet but needs 180 volts to do so and uses about 1,700 watts, or the European import—the **Sunrise** pump—which pumps as deep as 600 feet at 60 volts using only

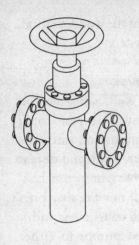

2 2 0 watts.

Why can one not use a normal well pump?

Because they run on either 120 volts or more likely on 240 volts. They use, depending on the depth of your well, about 1 to 2 hp, which translates roughly to 750 to 1,500 watts AC. If your well is not close to the house and is fairly deep, you might consider using an AC pump on either two inverters synchronized or a single inverter with a 240 volt transformer vs. a DC pump because, as you may remember, AC goes the distance easier than DC. AC pumps create a tremendous surge when started up because they initially have to work hard pushing up all the water that is in the line above

them. Their surge is higher than the the surge of a DC pump because their motors are stronger and capable of delivering much more force than a DC pump. They also usually deliver enough of a flow rate to pressurize your home.

DC pumps are useful because they are more efficient. If your well is too far away from the house, you can install a pump with an appropriate amount of solar panels at the location and let it pump as long as the sun shines (a *stand-alone system*). You would pump this water into a holding tank and either feed your house by gravity pressure or with the help of a DC booster pump.

Booster pumps can be as inexpensive as $95 or run up to $1,200, depending on your needs. They are usually used in conjunction with a pressure tank which maintains the pressure in the system until it is drawn down, which means the bladder inside the tank has retracted to a preset pressure. At this point, a pressure switch turns on the booster pump which tries to maintain the pressure in the line. A small pressure pump may be able to fill your pressure tank but may not be able to maintain the pressure in the water line, especially if more than one tap is turned on. This could mean that nobody else can use water while you shower. Bigger pressure pumps can overcome this problem and can maintain enough pressure in the lines to enable you to turn several taps on at once.

DC well pumps pump very little water at a time, usually 1 to 3 gallons per minute. That's why you need to have a holding tank to collect this water and a pressure pump to pressurize your system.

If you choose to collect water from rain only, and collect it in a cistern, all you need is a booster pump to pressurize your system. This is pretty common in Austra-

lia and the Mediterranean countries.

Now, here is an interesting story. It is called: WHO OWNS THE RAIN? Consider this: a normal American household is estimated to use up to 1 acre foot of water per year, which is about 360,000 gallons. An enormous amount, about 1,000 gallons per day. This is all nice, legal, and mostly free (if you have your own well). But beware! If you catch rainwater, let's say in the Southwest where there are about 12 inches of rain per year (depending on your roof size this translates to 12,000 to 24,000 gallons of water per year), this rain is **not** free. Legally, you are not entitled to catch this rain— it belongs to the government. Now, fortunately, as of yet, the government has not enforced its entitlement, but I was told it could do so at any time, if it so chooses. (Probably they will let those black helicopters from the UN fly over your house at night and prevent you from catching the rain.)

Catching rainwater, whether you have a well or not, is a good method if you live in arid areas, because the water you take out of the ground will be missed somewhere downstream. Filling your cistern with rainwater will relieve the demand on the underground water table and make your well pump work less and use a "free" (so far) resource.

GENERAL LIGHTING

I just got a phone call from a new customer. She

bought a bunch of inexpensive fluorescent lights but they have a hard time start-

ing up and won't stay on. (We talked about this earlier, but it is important to repeat because it happens over and over again.)

I explained to the customer that certain small fluorescent lights do not run well on inverters. Their wattage is too small to start up the inverter or they have what is called a *mechanical ballast*. As I described earlier, a fluorescent light, in contrast to an incandescent light, does not "burn." The electricity running through the ends of the tubes only triggers a process inside the tube, which is filled with a certain gas like neon, that makes the gas molecules start moving and colliding with each other. This process gives up a radiation of ultraviolet light which then collides with the phosphorous coating on the inner tube which creates visible light.

Hardly any energy is used in this process. Since this would mean that the electricity would create an almost direct short between the hot wire and the neutral, an artificial user is introduced which is called a *ballast*. There are two types of *ballasts, mechanical and electronic. Mechanical ballasts,* as used in conventional bigger fluorescent lights, have a harder time starting up on inverters and in cold weather. Compact fluorescent lights, at least of the newer type, have *electronic ballasts* which have no problems with inverters and cold weather.

These are the light bulbs of your choice when it comes to AC lighting. They come in different shapes, open short tubes or covered with plastic globes in different tones. They are good for general lighting since they spread their light evenly. They can also been used in recessed light fixtures (but you have to buy fixtures specially designed for them).

The main reason to use compact fluorescent light bulbs and not incandescent bulbs is the difference between wattage vs. the light they give up. Incandescent light bulbs burn inside. They have a high resistant metal filament heated up by electricity so it gives visible light. It also and foremost gives up heat, which is a waste product.

A 20 watt compact fluorescent light bulb gives as much light as a 75 watt incandescent bulb. Its life expectancy is about 10,000 hrs vs. 1,000 hrs for a normal light bulb. However, the price is much higher—about $15 to $25 because there is much more technology involved to build them. However, considering the longer life expectancy, they are about the same or even cheaper than normal light bulbs.

They are available in almost any electrical store (but watch out for the electronic ballast) and fit almost any light fixture, if you buy the right size and shape.

Halogen Lights

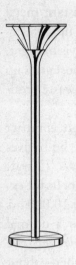

As already mentioned in the kitchen section, halogens are the perfect choice for DC lighting. They usually come as 12 volt units, often with a transformer attached. You can get them without a transformer or you can bypass it. They are named after two Swedish words *Halo* and *Gen* and use a small quantity of one of five non-metallic elements (for the curious: fluorine chlorine, bromine, iodine, and astatine) for a discharge within a vacuum tube or bulb. Their light efficiency is enormous, and you can uses a 35 watt bulb or even less as a good source for a reading lamp. As I mentioned earlier, there are rumors that they give up strong electromagnetic radiation (which is true), which might have some effects on people. (I used an electromagnetic field tester on a variety of sources and found that, at a distance of three feet, their discharge is much less than that of a 25 inch TV at a distance of six feet.)

Since halogen lights usually run on a 12 volt DC source, you can connect them either as single lights, parallel them, or, if you have a 24 volt or higher system, connect them in series.

GENERAL HEATING

The uniform building code requires that every house has to have a "proper" heating system. Wood stoves don't seem to qualify in most cases. This means that you have to install a proper heating system, designed according to the size of the house. So far I have not seen any evidence that a passive solar design will give you any credit toward the size of your heating system. This is just too bad because a properly designed passive solar home can—in certain regions like the Southwest—eliminate the need for any heating system. Designs like the Earthship or the passive solar shape developed by James Kachadorian (***The Passive Solar House***) are capable of keeping a constant room temperature all year round. Inspectors are often not qualified enough to follow the design specifications and calculations of passive solar homes in order to determine the right size for back-up heating.

Heating systems are expensive, and if you do not need them, those are wasted expenses. I came across three solutions to this problem which were successful in the state of New Mexico:

1. Install direct vented gas heaters around the house. They do not need any electrical assistance and the total bill will be considerably lower than for a cen-

tral heating system with zones and thermostats. The advantage is that you only have to turn them on when really needed or keep their thermostats on a setting that will au-

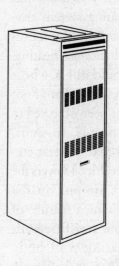

tomatically keep the room at a minimum temperature.

2. If you are 100% sure that you will not need any back-up heating, like the southern part of the Southwest, install electric baseboard heaters around the house (the cheapest solution) and provide a back-up generator that can run those heaters if needed. (Needless to say, no solar system can run any electric baseboard heaters.) You most likely will

153

never have to use them. They fulfill only the Code requirement.

3. If you think you need a heating system anyway, the latest movement is toward radiant floor heating. Now I could fill a whole chapter on why radiant floor heating has been objected to by many people. In short, they claim that the heat created is unhealthy. Flowers die when pots are put on the floor. Their nice feature of heating your feet first leads to overheating your body because by nature the feet are always the coolest part of your body. I cannot verify any of these claims—I just put them out for discussion. I only know that, in parts of Europe where this heating was fashionable for a while, the trend is moving away from radiant floor heating.

The problem that arises when using a heating system with zones and thermostats is again with our well-known phantom load. A radiant heating system has one or several circulation pumps and some electronic boards, as well as a transformer. All this will keep your inverter running on idle even when the system is on stand-by. Of course, there is a solution to this problem.

You need something that turns the whole system on when needed and off when not needed. Usually the boiler is on stand-by and when one of several thermostats switches on (because it got too cold in this room or zone), it turns the boiler and the circulation pumps on. Now the problem is that those thermostats run on 24 volts supplied by the circuitry of the boiler. If the whole boiler is off, how will the thermostat get its voltage?

Remember, you already have 12 or 24 volts in your batteries. What an incredible piece of luck, isn't it? You choose one thermostat as the master thermostat. Usually you will choose the one in the coldest spot of the house. You supply this thermostat with DC from your batteries. When it gets cold in this zone, the thermostat will close a connection and the voltage supplied from your battery will switch a small relay which, in effect, will turn on the power to your boiler. The boiler will awake from its dormant state, circulation pumps will start pumping, zone valves will open, and the house will heat up. If it gets too hot in the zone of the master thermostat, the connection will open, the relay will disengage, and the boiler will go back to sleep. Sounds simple, doesn't it?

Or to be even simpler, you can run the 110 volt power that will eventually turn on your boiler to a *line thermostat* which can switch on 110 volts. This is the "master thermostat." Once it calles for heat, it will switch on the boiler system.

But if your house is a passive solar design and built properly, you should not have to use too much for back-up heating. Radiant floor heating really does not qualify for back-up heating because it takes one or sometimes up to two days to heat up all the floors and let the warm floors heat up the house. A real back-up heater should be able to supply instant heat for only a short time. So if you decide to follow the latest trend of heating your floors, you might want to plan on having the heating system on for most of the winter, which can be an expensive choice.

INVERTERS

We already talked about what an inverter is. You know the definition—one that inverts! Now let's see what type of inverter you need and what brand is useful. There are cheap inverters that you see in hobby and tool mail order catalogues. Sometimes they give impressive figures for unbelievably low prices. Well, in those cases the saying is true—if it is too cheap to believe, better not believe it.

These inverters are what is called *solid state* inverters, i.e., no moving parts and no adjustments. They are on from the moment you plug them in until you unplug them. They might be good for short-term applications, as advertised, to run small tools off your car batteries. My experience, however, is that, in many cases, their advertised search wattage does not live up to its expectations in real life. I had an inverter with a 500 watt search which could not start a 19 inch TV. It did not run a small circular saw and barely managed a small electric drill.

Other inverters, which are designed for constant use and permanent installation, cost considerably more but they live up to their expectations. The top brands in conventional inverters are **Trace** and **Heart**. Both designs are solid and have been tested over the years. **Trace,**

being the leading inverter manufacturer, has a few drawbacks as far as service is concerned. Inverters rarely break. In fact, I have had only one inverter break or malfunction in over six years. It was my own. I sent it back to **Trace,** explaining that it was all I had to keep me happy. After a **month**, I called back to inquire about it and it turned out that they could not find it. After **six weeks**, I finally got my inverter back. I think this is a bit slow considering that it was the main power supply for a dwelling.

Now I have heard that they have beefed up their service a bit. But even I, as a dealer, do not get good technical support on the phone. It takes forever to get questions answered because every technician seems so specialized that he can only answer one question at a time and has to refer you to somebody else for further questions. I write this not to put the product down—it is still my number one choice of inverter—but if my customer suffers, I suffer, and a company this big should brush up on their service. Thank you.

The **Heart** inverters are very, very sturdy inverters. They never create any problems. **Heart** also has several contract repair stations nationwide, which assures you excellent service in your local area. These inverters come, as do the **Trace** inverters, with a search mode that turns the inverter to stand-by when it is not in use, as well as with battery chargers. Which means they automatically transfer over and charge your battery bank when they sense an incoming voltage. Considering that a good battery charger and a transfer switch can cost around $500, this is an excellent feature.

Of course, the **Trace** inverters are designed for all possible home power requirements. They come in sizes up to 10,000 watts.

They can accept several incoming sources, which makes them an ideal choice if you want to interface your system with the utility grid. They also can be programmed to start your generator and, last but not least, all of the bigger models (4,000 watts and up) are true sine wave inverters. (I forgot to mention that they make you coffee in the morning and take the kids to school.)

Another inverter brand is the **Exeltech** inverter. These are fine inverters for all kinds of electronic equipment because they deliver a clean true sine wave and come in small sizes. If you do not want to invest in a 4,000 watt inverter or have a system that cannot support one but still want to have a clean wave, this will be your choice. They also come as a modular design, which means you can start small and expand by adding more modules to your existing module without interrupting operations. However, the

Exeltech inverters are not as efficient as **Trace** and **Heart** inverters.

Another brand is just entering the race with smaller sine wave inverters (which definitely is the trend for the future): **StatPower Inverters**. They also come with battery chargers and stand-by function. They neatly fill the gap in small- and medium-sized sine wave inverters that **Trace** never ventured into.

SOLAR PANELS

A lot can be said about solar panels. Most of

it will be very technical and does not really help you to understand your system better. The biggest brands are **BP, Simens,** and **Solarex**, all

of which have their advantages and shortcomings. Again, any discussion of these advantages and shortcomings would be very detailed and technical and will not affect you very much once the panels are installed. Most solar panels, one can say, live forever, unless you have an automated spacecraft fly into them while trying to dock at your main entrance.

The issue is mostly whether to use recycled panels vs. new ones. It all depends. Older panels tend to use up more space because they have less power and you need to put up more panels to get the desired wattage. The most efficient panels are the **BP 590**, with 90 watts—very powerful. (Of course, technicians have a few words to say about the slightly higher output voltage which will not be reflected in usable watts.) Other panels like the **Simens 75 watt** panels are excellent and so are the **Solarex** panels. In the end, it comes down to your budget, and which brand is on sale at the moment. Of course, their size also matters at times. The new generation of flexible thin film panels, like **BP Apollo Modules**, are just entering the market. They are supposed to bring the price down considerably.

I have personally installed a lot of used panels over the years, which came out of recycled solar power plants in California, and they all performed well to their specifications. So mostly it is up to your solar designer which brand he/she feels comfortable working with. For example, some of the **BP** panels have a certain type of connector inside their junction boxes which is very hard to connect to. But this problem has now been corrected.

FUSES AND BREAKERS

Most fuses and circuit breakers are not rated for DC use, which means you cannot legally install them in your DC circuitry. They may not function. There are very few breakers which are DC rated. Fortunately, one of those is readily available in any hardware store. It's the **Square D** brand. Not all **Square D** breakers are rated for AC as well as for DC— only what is referred to as the **QO-breakers**, which are the original **Square D** breakers. **Square D** bought out a cheaper line called **Home Line**. Those breakers are not DC rated.

Another brand which you will find in solar mail order catalogs is called **Heineman** breakers. These breakers are normally used in more specialized applications, since they will not fit into any common breaker box. You will find them in **Trace's** main disconnect panels and power panels.

Fuses are a much more difficult issue. Almost none of the fuses available in electrical supply stores are DC rated. So mostly you find normal AC fuses inside fused disconnects. In most cases, they work just fine.

However, there is one kind of fuse which is DC rated and very inexpensive and readily available. This is the **Auto** fuse. In many older systems, you will find these as either glass or plastic fuses. These fuses are not UL listed and, of course, not approved by Code. Also I found that their design is rather weak as far as attachments and connections are concerned and have led to sparking and fire inside their housing. I would not use any of them in normal DC fusing applications. The only exception is when small lines such as feeder lines to metering devices are in-

volved. Here, we only need up to 5 amps max and an in-line-fuse will do the job adequately.

LIGHTENING PROTECTION

This is a somewhat difficult section because there is no adequate lightening protection. The word "protection" is misleading here. If you are lucky, you can avoid the worst.

People often ask whether or not to install a lightening protecting system. The answer is I DO NOT KNOW!

One customer had an elaborate lightening protecting system installed. He spent over

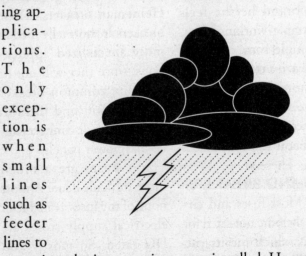

$4,000 on it. But almost every component in his house has been hit by lightening since. Nobody knows whether he attracts lightening or whether he has avoided the worst because of the system.

Besides those professional lightening protection systems, which should be installed by a licensed and insured contractor, there are some devices on the market which may avert the worst if you get hit by lightening or have a nearby hit. They are called *lightening arrestors*. Again there are two different types available. One type is connected in line and reacts so fast to lightening that it shuts down before the lightening can do any further harm to your system. They have to be replaced after a strike because they are destroyed. These are probably the safest devices you can use.

Another device functions like an absorbing device. It is good only for near-by strikes and may also go kaput when hit heavily. Both devices are of affordable cost and you can find them in any solar mail order catalog. Usually two are installed, one nearest the array and one nearest the power center. If you only have one device, I would install it near the array, where you would most likely get hit, in order to defuse some of the electrical load early on. But there is really no guarantee that any of these round little cylinders will save you in case of a direct hit. The best is, if you are at home, throw the disconnect switch which will disconnect the array from the rest of the house.

Chapter Eleven'
CLEAN ELECTRICITY
How Safe Is Our Electrical Environment?

I do not consider myself as belonging to those doomsday people who believe that Phoenix, Arizona will build a harbor soon and black helicopters will finally be visible in the dark nights over the desert. All I know is that, over the last 50 years or so, the amount of electrical wiring in houses has more than doubled. We are surrounded by invisible rays especially of the microwave category, that phone companies shoot horizontally across the ground, that make reception of TV satellite frequencies impossible when located in the vicinity of a microwave tower. Radar signals near airports function almost like x-rays and, if you try to get any AM Radio reception near power lines, you will be disappointed.

Our living environment certainly has changed 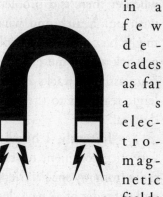 in a few decades as far as electromagnetic fields are concerned. And as proponents of all this high tech stuff are quick to point out, no study has yet shown that any of the above have damaged our bodies to the extent that we suffer considerable health problems.

Opponents will point out that, indeed, there have been studies in Sweden, Germany, and Switzerland and that there are villages in the U.S. where people near high tension lines have a disproportionately high rate of cancer.

This discussion reminds me somewhat of the discussion being held in Germany in the 1970s about the dying trees. The same Chancellor who later promoted himself as being the savior of "our" forests was the leader of the group denying that there was such a thing as dying trees. And while nobody could come up with conclusive studies whether or not acid rain caused any damage to the trees, whole forests (like the famous Black Forest) began to dwindle.

We know for a fact that people today are not as healthy as their parents and grandparents were. Diseases like allergies, chronic fatigue syndrome, constant colds, certain cancers, etc., are making people's lives rather miserable and creating great expense for everyone. The fact that more and more people seek alternative methods of healing shows that conventional medicine does not seem to be able to deal with those diseases effectively.

Of course, as stated earlier, there are no conclusive studies that show that any of these diseases are being caused by any of the many possible threads in the environment, like polluted food with hormones and pesticides (the estimated amount of foreign particles in a single conventionally grown tomato is 50), building materials (foremost, formaldehyde in carpets, walls, insulation, and furniture), clothes (bleached cotton contains an amount of bleach that causes skin irritation and pesticides from growing the cotton), drinking water (contains high amounts of chlorine, pesticides, and heavy metals), and, last but not least, electromagnetic fields.

Some people say that it might be safer living outside without a tent than inside our polluted houses.

As hinted at earlier, we do not know whether one factor causes any of the health problems we endure or whether it is an accumulation of several factors. But while politicians discuss whether there is a problem with our health and what might be causing it, people get sick and some even die from asthma attacks and allergic reactions to their underwear.

Recently, I heard a very nice comment on talk radio from my special friend and defender of all gays, lesbians, and environmental wackos, Rush Limbaugh. He was claiming that all this discussion about olestra, the non-fat oil, was totally unsubstantiated because he is eating the product on a regular basis and not experiencing any discomfort like diarrhea or stock problems.

Well, there it is, the proof of how safe all these products are. But when someone is blaming their cancer on living for decades under a high tension line, this is no proof that anything is wrong with electromagnetic fields!

Even if we do not know conclusively what causes what disease, would it not be smart, for the time being, to avoid as many of these possibilities that may cause problems with our health until scientists, paid by the right people, find out more about environmental hazards to our health? (I don't want to start about the tobacco industry executives who swore in affidavits that nicotine is not addictive and they were not increasing nicotine levels in cigarettes, and they are not targeting kids as consumers of their products. No, I don't want to talk about it.)

Now, let's for a minute assume that there might be the slightest risk that crossing a six lane freeway at 5 p.m. in New York, LA, or Chicago might be hazardous to your health.

Would you rather not attempt to do it? One would call that common sense. Of course, there is no study that yields conclusive results that there is a connection between severe health hazards from crossing a six lane freeway at 5 p.m. simply because nobody has done it.

Along those lines, let me mention an article in the May/June 1999 issue of **Sierra** magazine, entitled "Current Risks: Experts Finally Link Elecromagnetic Fields and Cancer." The article goes on to say, "For 20 years, scientists have been trying to determine whether a mainstay of modern life might be increasing your risk of cancer. Finally, last June, a panel convened by the National Institute of Environmental Health Sciences decided there was enough evidence to consider the invisible waves called electromagnetic fields, like those generated by power lines and electric appliances, a 'possible human carcinogen.' Yet with no federal funds on the horizon, the NIEHS is pulling the plug on more research."

As it seems that no one will do the necessary studies in the near future, let's just, for argument's sake, assume that there is a connection between electromagnetic fields and certain health risks like brain tumors, and other cancers. Would you not rather err on the safe side and try to eliminate or reduce those fields in your living environment? One would call that common sense.

If you look at the electrical layout of a normal family dwelling, you will discover that there is not a single wall that does not contain a myriad of electrical wires. But before we get involved in the design or misdesign of houses, let's discuss the term *electromagnetic field*.

You have been introduced to the terms and definitions of electricity and especially of alternating current or AC (***Chapter Three***,

p. 19). Whenever AC is created by coils and magnets and the frequent change of its polarity (60 Hertz), there is an electromagnetic field associated with the presence and in particular with the flow of AC current. Simply speaking, wherever there are cables that contain AC, there is an electromagnetic field around them. This is true for every electric motor and electronic device. The general rule is: the higher the current in those wires or devices, the greater the field.

We took some tests with an electromagnetic field tester inside a conventionally wired house equipped with a normal 150 amp single

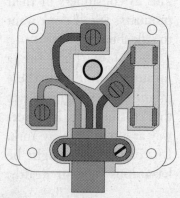

phase 220 volt service. We discovered several hot spots inside the walls of the house.

By far the most active part and the strongest field was detected near the breaker panel. As a matter of fact, one had to get three feet away from the box before our tester registered within the "normal" and less dangerous category. The same was true for the TV. A 25 inch color TV set required a minimum distance of four feet before the field was acceptable. (Of course, what is acceptable and what is dangerous was defined by the manufacturer of the tester. There are no standards for electromagnetic field testing yet.)

One of the most interesting and least suspected danger areas in the house was the master bedroom, in particular because this house was equipped with an electric baseboard heating system. It turned out that the wires to the heater were run around the walls of the bedroom and ended inside the heater on the opposite wall to the bed. When turned on, there was no place inside this bedroom that was not considered a "danger zone" by our tester. The way bedrooms are typically wired,

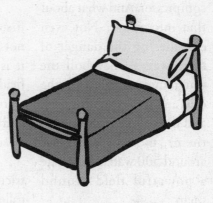

you find a receptacle on each side of the bed and often one or two lights right above the bed. This means that the wall where you rest your head for seven to eight hours of sleep is often at exactly the height to have it lit up like a Christmas tree by the electromagnetic field of the wires passing through the wall. Unless you have a headboard made of 1/4 inch steel, your nervous system will be enhanced by those fields (just in case you ever wondered why you have such hot dreams sometimes).

Needless to say, the kitchen is a hot area, too. Consider the refrigerator or freezer with its powerful compressor. And what about that microwave? Not even considering the danger of microwaves (they boil the water molecules inside the cells and so heat up the food and the people coming near them), most of them use around 500 watts and create a powerful field around them.

We can go on and on, passing by the electric hot water heater and several other appliances. You will argue that you usually don't spend that much time sitting next to your hot water heater or fridge. But as I stated earlier, the wires going to those appliances may pass through your living room walls right behind your favorite place near your nice big screen color TV to make sure you get "fried" front and back equally.

You may ask now what we can do to "clean up" a situation like that and what is the difference between conventional electricity and solar electricity.

The answer to the first part of the question is not an easy one even though it is possible to reduce the fields in your walls. But let's start with the second part and from there we may find a way to answer the first part

Conventional electricity is delivered at what is called 220 volts single phase. Which means that you get two conductors each carrying 110 volts in them delivering electricity to your house. These two 110 volt conductors are out of phase to each other and only between them do you get 220 volts, while you get 110 volts between each of them and ground.

Your electricians run wires containing either or both of these 110 volt phases throughout your house. Always depending on the user at the end of those lines of course, it can be said that, in general, the field around 220 volt lines are stronger than around 110 volt lines, simply because the tension between them is higher.

In a solar application, in most cases, we only deal with one single phase 110 line, which by nature has a weaker field than a 220 volt line. A typical solar house with a 4,000 watt inverter (which is considered pretty big) uses around 35 amps max. A typical residential dwelling has the potential for about 150 amps although typically the usage is around 70 to 80 amps, which is still twice as much as a solar house. Lately, houses are getting more powerful and 200 amp services are more common.

The biggest plus, however, is a feature in solar houses which is designed into the AC power inverter. This feature is a *stand-by function* that turns the inverter off into a stand-by mode when no electricity is used. Now, a great deal of headache goes into the design of a solar home with regard to this feature. A lot of compromises have to be made to make use of this function but, if properly designed, the house will be "off" at night and no electricity will create electromagnetic fields inside your bedroom walls or elsewhere in the house.

Another feature of a solar home is its capability of using DC, which is direct current straight off the battery bank. Many lights can be run off DC and so the amount of AC current and its increased electromagnetic field can be reduced.

You may ask now: "How can I save my hopelessly mis-designed conventionally wired house?" Or: "What can I do if I have 'grid power' available and still want to design a 'clean' house?"

First of all, you have to get rid of any electric heating in the house. Besides the fact that you waste a lot of money on your electric bill, you need to replace these big users with gas-powered appliances. But even then, you are still stuck with your conventional design and power inside your walls at night.

If you reduce the amount of electrical current at night, you help some, but there is, of course, the possibility of creating a hybrid system, a mixture between grid power and solar power. Even though still rare, those systems exist and some of them are very successful. You can always power parts of your house with solar energy and other parts with conventional grid power. However, considering the prices of solar systems, you will not profit from it in the short run. But if your health is worth it to you and you consider the medical bills for some chronic diseases, it may, in the long run, be a viable solution. (Remember the discussion about *net metering* and combined solar and grid power in **Chapter Six**, pp. 38-39?)

The feedback from several very environmentally sensitive people to the feeling of solar electricity was very positive. In all, they felt the house was quieter, they felt less "caged in," and they could sleep better and deeper. Even if these statements are scientifically very subjective and not measurable observations, the results they deliver are more than worth it for those people.

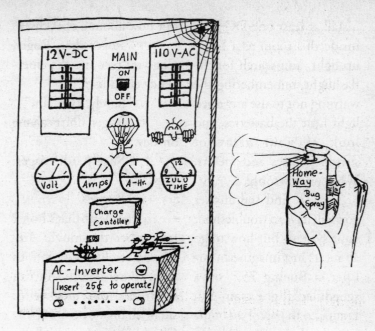

Chapter Twelve
TROUBLESHOOTING
Looking for Bugs

The sun is setting and silence is descending with the veil of darkness upon the lonely desert. Happy and content, you turn around and walk back into your newly equipped house and, in anticipation of a quiet evening watching your favorite TV show (which you had to hold off on for many weeks), you flip the light switch on.

Silence and darkness are all you experience inside as well as outside the house. The AC lights do not come on. You walk into the utility room and notice that it would have been smart to install at least one DC light inside this room as a back-up light. You search for the flashlight, remembering the warning not to use any open light near the batteries, and look at your array of switches, fuses and meters. What went wrong?

To find the answer you will have to troubleshoot your system. But how to go about it? Just imagine an airliner, a Boeing 757, on a scheduled flight from San Francisco to JFK. Two hours into the flight, the red lights come on inside the cockpit. A buzzer sounds, a bell rings, and the on-board computer flashes the message: "You have been terminated! Insert 25 cents for another game!" The captain remains cool. He first starts to count the four stripes on his uniform, then reaches into his pocket for change, but finds none. Then, he turns to his first officer.

"Roger, do you have any change on you?"

"Negative, Captain, but how about a checklist?"

"Roger that, Roger. Can you pass me the emergency checklist?"

"Surely, Captain."

"Roger, don't call me Shirley."

"Surely, uh, roger, Captain."

"Roger!"

(At least that's how I remember the movie *Airplane*.) You have two options in an emergency. You can reach into your pocket for small change or call for the Checklist. Since you will not find a slot in your system to insert a quarter for another game, all you can do is toss a coin or go directly to option two.

Option two, as we recall, is troubleshooting with the help of a checklist or without one, by trying to go through your system in an orderly manner.

If you had installed a DC light in your utility room, you would instantly know whether you still have DC power, which would

172

shorten your troubleshooting path. Let's assume (just for argument's sake) that you did install a DC light inside your utility room. The light is still burning, which indicates that your source for DC power is still functioning, at least in this room. Try to remember whether there are other DC sources in the house like DC lights, an answering machine, a radio, or a tape recorder. Go and see if they are working. If they are, you know now that you have an AC problem.

The general rule for all troubleshooting is to start at the source and work your way downstream. The source of your AC power, Point Zero, is the inverter. Go to it! Check the ON/OFF button and reset it. In 90% of all cases, this will restore your power and your peace of mind because on many inverters this switch doubles as a reset button for many overload or under-current situations. If everything fails, you can read the instructions on your inverter. Go to the section entitled *Troubleshooting* and follow the guide. (If you own an expensive **Trace** inverter, you will find that there is no indication of where the troubleshooting section is. Which is very helpful. If you find it eventually, you will also find that it is very confusing and of limited help.)

This was the troubleshooting guide for beginners. Any flight attendant or even any passenger with common sense could have saved this flight from crashing. But, by now, you have advanced to at least the first officer's position, and you can take on a more severe task. "Roger!" Now you have to find out

why this disturbance occurred in the first place.

And here is the new scenario:

There is no DC light in the house yet, the answering machine has not been delivered, and you are in total darkness. Now, remember, the lights may be off, but you are home. And you were smart enough to buy your yourself a small digital voltmeter at a nearby WalMart or Radio Shack.

Equipped with your voltmeter and a flashlight, you start at Point Zero.

Where is Point Zero? It is the source of your electricity. Remember, it is dark outside and your charge controller has turned itself off, so no power is coming in from the array. Your power is now coming from the battery

bank. This is Point Zero.

Of course, now is not a good time to read the instructions for the use of your voltmeter, therefore you have to use common sense. Your voltmeter has one or two dials. Turn one dial to a position that is labeled DC. (You remember what DC means, right?) And for the second dial, choose a range that your system voltage falls in. Let's say you have DC and numbers 3, 30, 300. Your system is 12 volts. The number to choose for the second dial is: 30! (Because 30 means that, in this DC section of the voltmeter, it will read up to 30 volts.) Of course, a lot of simple voltmeters are automatic and all you have to choose from is DC. Now we are ready to save this airplane from crashing!

We are now at Point Zero, the battery bank. It is dark and drafty down here. You will find two big wires leaving the battery bank. Before you follow them, make sure that you are not being followed. Then take the two probes which are attached to your voltmeter, one is red and one is black, and put one at the beginning of one big wire and the other at the terminus of the other big wire.

Check the digital readout of your voltmeter for voltage. If you have voltage, at least 12.0 or more (or 24 or more), you can now carefully follow the two big wires. They will either go to (1) the inverter or (2) a single fuse and then they may disappear inside a big box, the DC load center.

In case number one, make sure you have voltage at the inverter. If one cable goes through a single fuse (case number two), put one probe of your voltmeter first on one end of the fuse and the other probe at the nearest bare point of the other cable (the other pole of the battery, usually the minus pole). (You will not get a reading if you put them on both ends of the fuse.) Then do the same with the other side of the fuse. If you have voltage on both sides of the fuse, keep following the cable. If the two big wires disappear inside a DC load center, check for metering devices and disconnects on the box.

Try to find an indication for battery voltage and see whether it is high enough to run the inverter. Inverters often are low-voltage-protected, which means they turn themselves off if the voltage coming from the batteries is below a pre-set level. Usually 12.0 or 24.0 volts is enough to keep them running. But remember, if you operate some heavy equipment, like a big TV or heavy motors, the voltage may drop fast below the cutoff point of the inverter and, even though inverters usually have some time delay, it may have shut off because of the temporarily low voltage condition. So what you read may not be what you've got. If you wait a few minutes, the battery bank may have recovered to a sufficiently high voltage to start up the inverter again.

This big box most likely also contains a main fuse which, when blown, cuts off the rest of the system. If it is a pull-type fuse, **DO NOT PULL IT OUT!** If the fuse was not the problem, pulling it will erase any special programming in your metering system. Instead, take your voltmeter and check to see if there is voltage at the plus and minus sides of your inverter. If there is, try to restart the inverter by either pushing the ON/ OFF button or any other reset button that is labeled as such.

If you do not have voltage at the inverter, most likely the main fuse is blown. **Now stop and think!** Did you use any heavy equipment other than a back hoe, a steam roller, or a World War II tank recently? Could it have blown the main fuse? If your answer is "Yes," and you have a spare fuse, pull the main fuse and check it with your voltmeter.

To check a fuse once it is pulled out of the system, you have to set your meter to a different function. Turn the first dial to the position that reads *Ohm* and choose the lowest number on the second dial. Place your probes on both ends of the fuse and observe the meter. If the read-out goes to zero, the fuse is good. If nothing changes, the fuse has no continuity and is bad. Your voltmeter may have a function to do continuity checks. If you use this function, a beep will sound when there is continuity.

If you have a spare

fuse (which most likely you don't), replace it. (Please do not try to substitute the fuse with any other metal object such as nails, heavy wire, screwdrivers and such. If a severe problem caused this fuse to blow, you may create a fire by substituting a big piece of metal.)

If you replaced the fuse with its proper value and before you reconnect anything, **stop!** Make sure everything is turned off, except your flashlight. Make sure that every gadget is unplugged.

Now reconnect the fuse, and check whether the system is back on line. If it blows again right away, prepare for an evening with candlelight and leave a message with your electrician. You most certainly have a bigger problem than you can handle. But if it stays on, don't just celebrate. **Think again!** What could have blown the fuse? Did you get struck by lightening? (Or rather your house or the nearby area?) Did you use heavy equipment, except the above-mentioned? Could your battery voltage be too low and, if yes, why?

If you do not come up with a conclusive answer, start turning things on, one by one, and each time give it a few minutes to settle before you continue. In between, check your voltage for trends. Is it going down rapidly? Did it go down more than it should, considering the load? Consider the possibility that two heavy loads like a DC fridge and a water pump, which both turn themselves on automatically, came on at the same time and caused a momentary overload. Try to recall your last steps and actions as if you were the main suspect in the hot TV-show: *"Mystery, He Screamed!"*

Now here is another scenario. In the last few days, you have noticed that, after the end of the day, the battery voltage is not as high as it used to be, even though you have not been home during the day. As a matter of fact, on the third day, it indicated only 11.8 volts.

You do not remember leaving anything on while you were gone except may be the answering machine. What caused this? The answer is that, most likely, your batteries are not being charged.

Again, **start at the source**! Find the box or switch which disconnects your solar array. If it is still daylight, use your voltmeter and check the incoming voltage. You either have a separate disconnect with fuses inside, or it is all part of the big box called the *DC control center.*

If the latter is the case, search for a switch labeled ARRAY DISCONNECT and try to reset it. If the former is the case, open the switch box, find the fuses, and put the probes of your voltmeter at the plus and the minus of the incom-

ing line before the fuses and after the fuses. (At this point you do not really care whether you put plus to plus or minus to minus. The indicated voltage may or may not have a minus in front of the numbers, but the numbers will be the same regardless.)

If your voltage before the fuses is much higher than the voltage after the fuses, you know that one or

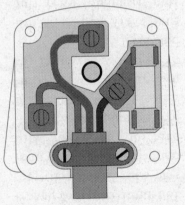

both of the fuses is bad. Pull the fuses out and check them with the ohmmeter of your tester. (Follow the above instructions on how to use the ohm scale correctly.)

If you discovered all of this after dark, you will not have any incoming voltage because your charge controller has turned itself off. Now you can check the fuses with the ohmmeter only. You can even do this without taking them out of the fuse holder. Just set your voltmeter to the ohm scale and put the two probes on both sides of the fuse. If you have continuity, check the other fuse.

In many cases, there is another disconnect right by the solar array. This may involve a climb up on the roof. Please, unless you are an experienced climber and are current for night climbing, postpone this until daylight. The advantage at daylight is also that you can check exactly up to what point you have charge power.

The charge power, if disconnected from the rest of the system, is also called *open circuit voltage*, which means the voltage your array puts out without anything connected to it, which is usually considerably higher than the battery voltage. The moment it is connected to a load or to a battery bank, it will drop down drastically, finding a value based on the condition of the battery bank.

So if you find that you have a much higher than expected voltage coming in from your array, follow it downstream until you find the point where it suddenly drops down to the lower voltage of your battery bank. If a fuse is between those two points, check the fuse. If the charge controller is between those points, you may have a problem here.

Some charge controllers also have reset buttons or fuses, some of them inside the controller itself. You can take the cover off and check the fuse by removing it. (Now we are talking advanced troubleshooting).

In this chapter, we learned that the general idea of troubleshooting is that, if you follow a logical sequence of steps, starting from the source and moving downstream, you sooner or later will come across the problem, which, in most cases, is a reset button or a blown fuse. To make things easier for you and, at the same time to make you feel like First Officer Roger or Captain Shirley, I will include some Checklists to follow. These are generic checklists and you may have to customize them to your system. But

t h e y
w i l l
g i v e
you a
start-
i n g
point,
which
i s
m o r e
than a
lot of

people have on a dark night in a dark house with only coyotes howling outside.

Checklists

CHECKLIST A:
NO AC POWER

1. Go to the source of AC power, the inverter.

2. Check if it is ON. If not, turn it back on.

3. If that doesn't work, check for any LED lights on the inverter and read the instructions on what they do and how to restore power.

With most **Trace** inverters, switching the power OFF and ON again will reset the inverter.

4. If no lights are visible and there are no other displays or indications and no power, check incoming voltage. (Two heavy-duty wires labeled plus and minus). If there is no voltage, you have a DC problem.

5. If you have incoming voltage and nothing resets the inverter, the inverter is the problem. Read the instructions or call the manufacturer.

6. If the inverter is ON or could be reset and you still have no AC power, check the AC load center. Check the main breaker and, if it is OFF, reset.

7. If all breakers check OK, the inverter is ON and there is still no power at any and all locations, you have a real problem. Call for help.

8. There usually is a big fuse between the battery plus and the inverter plus. Check it for continuity.

CHECKLIST B:
NO DC POWER

1. Verify that you have no DC power at all points.

2. Verify that you also have no AC power.

3. If you have no DC but you have AC, go to the DC load center and check for tripped breakers or blown fuses.

4. If everything checks OK there, open the breaker box and check for incoming voltage and trace it backward until you find a blown fuse or other problem. Always check for tight connections.

5. If you have no DC and no AC, go to the source, the batteries.

6. Check voltage and connections at all wires leaving the batteries.

7. Follow the wires and check all connections and fuses along the line.

8. Remember to always check voltage between plus and minus.

9. Remember to also check for proper settings on your meter.

10. If you can't find the problem, relax, light a candle and plan for a romantic evening. It might be a more difficult problem.

CHECKLIST C:
NO CHARGING

1. Go to the source, the array. If this is impractical or dangerous because the roof is wet or it is dark, find the ARRAY DIS-CONNECT.

2. Go to the point labeled ARRAY DISCONNECT.

3. Check for tripped breakers and reset them or replace blown fuses.

4. If you have no success, check the voltage on both sides of the breakers or fuses.

5. Remember, always check between the plus and the minus pole. (If you check between two pluses, you will not get any voltage indications.)

6. If the voltage on one side is much higher than on the other side of a fuse, pull out the fuse and test it for continuity. Before replacing a bad fuse, disconnect all power sources in the house.

7. If you had to turn off a disconnect to open a breaker box, make sure to turn it back on during testing. But turn it off when replacing fuses.

8. If there is no incoming voltage, start to follow the line backward toward the array. There may be other disconnects between the array and the house.

9. If there is no power at the next disconnect, check the array for loose connec-tions. (Pay attention to old wire nuts—they may be corroded or may not be twisted properly.)

10. If there are no visible loose connections but there is no power coming from the array, you need to call for help. One of your panels may be bad or improperly connected.

CHECKLIST D:
FOR POWER CENTERS

If your system is wired through and controlled by an integrated *DC Power center*, use the following procedure:

A. NO AC POWER

1. Check the inverter for power. LED lights or a digital LED light indicate an overload. Reset the inverter.

2. If the inverter indicates a low-voltage situation, resetting may not be possible and/or not advisable.

3. If you receive a low-voltage indication, investigate the reason for it. Are you charging properly?

5. If you think you have used too much power, recharge the batteries or use alternative power until enough power is restored to run the inverter.

6. If you think that you have not used enough energy to cause a low-voltage-situation, you may have a DC power problem.

B. NO DC POWER

1. Each power center usually has a main DC-disconnect. Either reset the disconnect or, if it is a pull-type fuse, pull it and check its continuity with a tester.

2. If the main disconnect is OK, check the array disconnect.

3. If the ARRAY DIS-CONNECT is OK and you still have no power, check to see if there is a disconnect located at the array.

4. If you recently had lightening activity near your house, it may have blown the lightening arresters installed in your system.

5. If you have a multi-

meter installed in your panel, check for:

    a) array voltage/ charge voltage;
    b) battery voltage;
    c) array amps;
    d) load amps.

6. If you have no array indication, you are not charging. (Of course, you can check the array only during business hours, which is during sun-hours).

7. If you are charging but the batteries are low, check the load amps with all loads turned off (use a flashlight). You should have no load amps indicated. If you do, you have a hidden load. Go and find it. You may have a short in your system.

8. If you cannot find any indication of a problem, consider the possibility of an intermittent problem like a loose connection.

9. Checking for loose connections can be danger-ous. (Sparks can cause arcing and melt connec-tors.) Make sure all power is off before you touch any connections.

A general warning: When working with DC wiring, especially loose con-nections, be careful. Even though you work with low voltage and you may not get an electrical shock, you may get a shock of a different kind. DC connections, when loose, tend to arc and spark strongly and can easily set things on fire or melt con-nectors. They can cause se-vere burns on your skin, too. If you discover any sparking when touching connections, turn the power off immedi-ately. Repair the problem and turn the power back on. Never attempt to fix a spark-ing connection while the power is still on.

APPENDIXES

APPENDIX A
Solar Water Pumping:
A Practical Introduction
by Windy Dankoff

If you need to supply water beyond reach of power lines, then solar power can solve the problem. Photovoltaic powered pumps provide a welcome alternative to fuel-burning engines, windmills, and hand pumps. Thousands of solar pumps are working throughout the world. They produce best during sunny weather, when the need for water is greatest.

How It Works

Photovoltaic (PV) panels produce electricity from sunlight using silicon cells, with no moving parts. They have been mass-produced since 1979. They are so reliable that most manufacturers give a 10-year warranty, and a life expectancy beyond 20 years. They work well in cold or hot weather.

Solar water pumps are specially designed to utilize DC electric power from photovoltaic panels. They must work during low light conditions at reduced power, without stalling or overheating. Low volume pumps use *positive displacement* (volumetric) mechanisms which seal water in cavities and force it upward. *Lift capacity* is maintained even while pumping slowly. These mechanisms include *diaphragm, vane,* and *piston pumps.* These differ from a conventional centrifugal pump that needs to spin fast to work efficiently. Centrifugal pumps are used where higher volumes are required.

A *surface pump* is one that is mounted at ground level. A *submersible pump* is one that is lowered into the water. Most deep wells use submersible pumps.

A *pump controller* (current booster) is an electronic device used with most solar pumps. It acts like an automatic transmission, helping the pump to start and not to stall in weak sunlight.

A *solar tracker* may be used to tilt the PV array as the sun moves accross the sky. This increases daily energy gain by as much as 55%. With more hours of peak sun, a smaller pump and power system may be used, thus reducing overall cost. Tracking works best in clear sunny weather. It is less effective in cloudy climates and on short winter days.

Storage is important. Three to ten days' storage may be required, depending on climate and water usage. Most systems use water storage rather than batteries, for simplicity and economy. A float switch can turn the pump off when the water tank fills, to prevent overflow.

Compared with windmills, solar pumps are less expensive, and much easier to install and maintain. They provide a more consistent supply of water. They can be installed in valleys and wooded areas where wind exposure is poor. A PV array may be placed some distance away from the pump itself, even several hundred feet (100 m) away.

What Is It Used For?
Livestock Watering: Cattle ranchers in the Americas, Australia, and Southern Africa are enthusiastic solar pump users. Their water sources are scattered over vast rangeland where power

lines are few, and costs of transport and maintenance are high. Some ranchers use solar pumps to distribute water through several miles (over 5 km) of pipelines. Others use portable systems, moving them from one water source to another.

Irrigation: Solar pumps are used on small farms, orchards, vineyards and gardens. It is most economical to pump PV array-direct (without battery), store water in a tank, and distribute it by gravity flow. Where pressurizing is required, storage batteries stabilize the voltage for consistent flow and distribution, and may eliminate the need for a storage tank.

Domestic Water: Solar pumps are used for private homes, villages, medical clinics, etc. A water pump can be powered by its own PV array, or by a main system that powers lights

and appliances. An elevated storage tank may be used, or a second pump called a booster pump can provide water pressure. Or, the main battery system can provide storage instead of a tank. Rain catchment can supplement solar pumping when sunshine is scarce. To design a system, it helps to view the whole picture and consider all the resources.

Thinking Small
There are no limits as to how large solar pumps can be built. But, they tend to be most competitive in small installations where combustion engines are least economical. The smallest solar pumps require less than 150 watts, and can lift water from depths exceeding 200 feet (65 m) at 1.5 gallons (5.7 liters) per minute. You may be surprised by the performance of such a small system. In a 10-hour sunny day, it can lift 900 gallons (3,400 liters). That's

enough to supply several families, or 30 head of cattle, or 40 fruit trees!

Slow solar pumping lets us utilize low-yield water sources. It also reduces the cost of long pipelines, since small-sized pipe may be used. The length of piping has little bearing on the energy required to pump, so water can be pushed over great distances at low cost. Small solar pumps may be installed without heavy equipment or special skills.

The most effective way to minimize the cost of solar pumping is to minimize water demand through conservation. Drip irrigation, for example, may reduce consumption to less than half that of traditional methods. In homes, low water toilets can reduce total domestic use by half. Water efficiency is a primary consideration in solar pumping economics.

A Careful Design Approach

When a generator or utility mains are present, we use a relatively large pump and turn it on only as needed. With solar pumping, we don't have this luxury. Photovoltaic panels are expensive, so we must size our systems carefully. It is like fitting a suit of clothes; you need all the measurements.

Here is a guide to the data that you will need to determine feasibility, to design a system, or to request a quote from a supplier.

Solar Pump Design Questionnaire

Next, we will determine whether a submersible pump or a surface pump is best. This is based on the nature of the water source. Submersible pumps are suited both to deep well and to surface water sources. Surface pumps can only draw water from about 20 feet (3m) below ground level, but they can push it far uphill. Where a surface pump is feasible, it is less expensive than a submersible, and a greater variety is available.

Now, we need to determine the flow rate required. Here is the equation, in the simplest terms:

Gallons (Cubic Meters) per Hour = Gallons (Cubic Meters) Per Day / Available Peak Sun Hours per Day. *Peak Sun Hours* refers to the average equivalent hours of full-sun energy received per day. It varies with the location and the season. For example, the arid central-western USA averages 7 peak hours in summer, and dips to 4.5 peak hours in mid-winter. (See *Appendix F*, p. 208.)

Next, refer to our performance charts for the type of pump that is appropri-ate. They will specify the size and configuration (voltage) of solar array necessary to run the pump.

Copyright © 1999 by Dankoff Solar Products, Inc. 2810 Industrial Rd. Santa Fe, NM 87505 USA (505) 473-3830 FAX (505) 473-3830 pumps@dankoffsolar.com www.dankoffsolar.com

APPENDIX B
Glossary of Solar and Water Pumping Terms

Courtesy of Dankoff Solar Products, Inc.
Revised by ADI Solar

Note: This glossary contains sections specifically for pumps and wells. Dankoff Solar made this available to me and I kept those sections because they interface with solar electric.

AC—ALTERNATING CURRENT: The standard form of electrical current supplied by the utility grid and by most fuel-powered generators. The polarity (and therefore the direction of current) alternates. In the U.S.A., standard voltages are 115V and 230V. Standards vary in different countries. Also see *Inverter.*

AMPS, AMPERE: The unit of measuring electrical current. Can be compared to the flow rate of water in pipes.

AMP-HOURS: Unit to measure the amount of amps used per hour.

ARRAY: Several solar panels arranged together, either in series or parallel. See *PV.*

CONVERTER: An electronic device for DC power that steps up voltage and steps down current proportionally (or vice-versa). Electrical analogy applied to AC: See *Transformer*. Mechanical analogy: gears or belt drive.

CURRENT: The rate at which electricity flows through a circuit, to transfer energy. Measured in *Amperes*, commonly called *Amps*. Analogy: Flow rate in a water pipe.

DC—DIRECT CURRENT: The type of power produced by photovoltaic panels and by storage batteries. The current flows in one direction and polarity is fixed, defined as positive (+) and negative (-). Nominal system voltage may be anywhere from 12 to 180V. See *Voltage, Nominal.*

EFFICIENCY: The percentage of power that gets converted to useful work. Example: An electric pump that is 60% efficient converts 60% of the input energy into work, pumping water. The remaining 40% becomes waste heat.

ENERGY: The product of power and time, measured in *Watt-Hours.* 1,000 Watt-Hours = 1 *Kilowatt-Hour* (abbreviation: **KWH**). Variation: the product of current and time is *Ampere-Hours*, also called *Amp-Hours* (abbreviation: **AH**). 1,000 watts consumed for 1 hour = 1 KWH. See *Power.*

GRID POWER: Electrical power as supplied by the utility company. See *Utility Grid.*

INVERTER: An electronic device that converts low voltage DC to high voltage AC power. In solar electric systems, an inverter may take the 12, 24, or 48 volts DC and convert it to 115 or 230 volts AC, conventional household power.

OPEN CIRCUIT VOLTAGE: Voltage of a solar panel (PV module) with nothing connected to it. The open circuit voltage of a 12 volt panel is typically between 17 and 20 volts.

PHOTOVOLTAIC: The phenomenon of converting light to electric power. Photo = light, Volt = electricity. Abbreviation: **PV.**

POWER: The rate at which work is done. It is the product of *Voltage* times *Current*, measured in *Watts*. 1,000 Watts = 1 Kilowatt. An electric motor requires approximately 1 Kilowatt per Horsepower (after typical efficiency losses). 1 Kilowatt for 1 Hour = 1 Kilowatt-Hour (*KWH*).

PV: The common abbreviation for photovoltaic.

PV ARRAY: A group of PV (photovoltaic) modules (also called panels) arranged to produce the voltage and power desired.

PV ARRAY DIRECT: The use of electric power directly from a photovoltaic array, without storage batteries to store or stabilize it. Most solar water pumps work this way, utilizing a tank to store water.

PV CELL: The individual photovoltaic device. The most common PV modules are made with 33 to 36 silicon cells each producing 1/2 volt.

PV MODULE: An assembly of PV cells framed into a weatherproof unit. Commonly called a "PV panel." See *PV Array*.

SOLAR TRACKER: A mounting rack for a PV array that automatically tilts to follow the daily path of the sun through the sky. A "tracking array" will produce more energy through the course of the day than a "fixed array" (non-tracking), particularly during the long days of summer.

STAND-ALONE SYS-

TEM: Refers to a solar water pumping system feeding a solar well pump directly by the sun without the use of a battery bank.

TRANSFORMER: An electrical device that steps up voltage and steps down current proportionally (or vice-versa). Transformers work with AC only. For DC, see *Converter*. Mechanical analogy: gears or belt drive.

UTILITY GRID: Commercial electric power distribution system. Synonym: mains.

VOLTAGE: The measurement of electrical potential. Analogy: Pressure in a water pipe.

VOLTAGE DROP: Loss of voltage (electrical pressure) caused by the resistance in wire and electrical devices. Proper wire sizing will minimize voltage drop, particularly over long distances. Voltage drop is determined by four factors: wire size, current (amps), voltage, and length of wire. It is determined by a consulting wire sizing chart or formula available in various reference tests. It is expressed as a percentage. Water analogy: Friction Loss in pipe.

VOLTAGE, NOMINAL: A way of naming a range of voltage to a standard. Example: A "12 Volt Nominal" system may operate in the range of 11 to 15 Volts. We call it "12 Volts" for simplicity.

VOLTAGE, OPEN CIRCUIT: The voltage of a PV module or array with no load (when it is disconnected). A "12 Volt Nominal" PV module will produce about 20 Volts open circuit. Abbreviation: **Voc.**

VOLTAGE, PEAK POWER POINT: The

voltage at which a photovoltaic module or array transfers the greatest amount of power (watts). A "12 Volt Nominal" A PV module will typically have a peak power voltage of around 17 volts. A PV array-direct solar pump should reach this voltage in full sun conditions. In a higher voltage array, it will be a multiple of this voltage. Abbreviation: **Vpp**.

PUMPS AND RELATED COMPONENTS

BOOSTER PUMP: A surface pump used to increase pressure in a water line, or to pull from a storage tank and pressurize a water system. See *Surface Pump*.

CENTRIFUGAL PUMP: A pumping mechanism that spins water by means of an "impeller." Water is pushed out by centrifugal

force. See also *Multi-stage*.

CHECK VALVE: A valve that allows water to flow one way but not the other.

DIAPHRAGM PUMP: A type of pump in which water is drawn in and forced out of one or more chambers, by a flexible diaphragm. Check valves let water into and out of each chamber.

FOOT VALVE: A check valve placed in the water source below a surface pump. It prevents water from flowing back down the pipe and "losing prime." See *Check Valve* and *Priming*.

POSITIVE DISPLACEMENT PUMP: Any mechanism that seals water in a chamber, then forces it out by reducing the volume of the chamber. Examples: piston (including jack), diaphragm, rotary vane. Used for low volume and high lift. Contrast with centrifugal. Synonyms: volumetric pump, force pump.

IMPELLER: See *Centrifugal Pump*

JET PUMP: A surface-mounted centrifugal pump that uses an "ejector" (venturi) device to augment its suction capacity. In a "deep well jet pump," the ejector is down in the well, to assist the pump in overcoming the limitations of suction. (Some water is diverted back down the well, causing an increase in energy use.)

MULTI-STAGE CEN-TRIFUGAL: A centrifugal pump with more than one impeller and chamber, stacked in a sequence to produce higher pressure. Conventional AC deep well submersible pumps and higher power solar submersibles work this way.

PRIMIMG: The process of hand-filling the suction pipe and intake of a surface pump. Priming is generally necessary when a pump must be located above the water source. A self-priming pump is able to draw some air suction in order to prime itself, at least in theory. See *Foot Valve.*

PULSATION DAMPER: A device that absorbs and releases pulsations in flow produced by a piston or diaphragm pump. Consists of a chamber with air trapped within it.

PUMP JACK: A deep well piston pump. The piston and cylinder is submerged in the well water and actuated by a rod inside the drop pipe, powered by a motor at the surface. This is an old-fashioned system that is still used for extremely deep wells, including solar pumps as deep as 1000 feet.

SEALED PISTON PUMP: See *Positive Displacement Pump.* This is a type of pump recently developed for solar submersibles. The pistons have a very short stroke, allowing the use of flexible gaskets to seal water out of an oil-filled mechanism.

SELF-PRIMING PUMP: See *Priming.*

SUBMERSIBLE PUMP: A motor/pump combination designed to be placed entirely below the water surface.

SURFACE PUMP: A pump that is not submersible. It must be placed no more than about 20 ft. above the surface of the water in the well. See *Priming.* (*Exception:* See *Jet Pump.*)

VANE PUMP—Rotary Vane: A positive displacement mechanism used in low volume high lift surface pumps and booster pumps. Durable and efficient, but requires cleanly filtered water due to its mechanical precision.

SOLAR PUMP COMPONENTS

DC MOTOR, BRUSH TYPE DC: The traditional DC motor, in which small carbon blocks called "brushes" conduct current into the spinning portion of the motor. They are used in DC surface pumps and also in some DC submersible pumps. Brushes naturally wear down after years of use, and may be easily replaced.

DC MOTOR, BRUSHLESS: High-technology motor used in centrifugal-type DC submersibles. The motor is filled with oil, to keep water out. An electronic system is used to precisely alternate the current, causing the motor to spin.

**DC MOTOR, PERMA-
NENT MAGNET:** All DC
solar pumps use this type of
motor in some form.
Being a variable speed
motor by nature, reduced
voltage (in low sun) pro-
duces proportionally
reduced speed, and causes
no harm to the motor.
Contrast: induction motor.

**INDUCTION MOTOR
(AC):** The type of electric
motor used in conventional
AC water pumps. It
requires a high surge of
current to start and a stable
voltage supply, making it
relatively expensive to run
from by solar power. See
Inverter.

**LINEAR CURRENT
BOOSTER:** See *Pump
Controller* . *Note:* Al-
though this term has
become generic, its abbreviation "LCB," is a trade-
mark of Bobier Electronics.

PUMP CONTROLLER:
An electronic device which
varies the voltage and
current of a PV array to
match the needs of an
array-direct pump. It
allows the pump to start
and to run under low sun
conditions without stalling.
Electrical analogy: variable
transformer. Mechanical
analogy: automatic trans-
mission. See *Linear Cur-
rent Booster.*

WATER WELL
COMPONENTS

BOREHOLE: Synonym
for drilled well, especially
outside of North America.

CASTING: Plastic or steel
tube that is permanently
inserted in the well after
drilling. Its size is specified
according to its inside
diameter.

CABLE SPLICE: A joint
in electrical cable. A
submersible splice is made

using special materials available in kit form.

DROP PIPE: The pipe that carries water from a pump in a well up to the surface.

PERFORATIONS: Slits cut into the well casing to allow groundwater to enter. May be located at more than one level, to coincide with water-bearing strata in the earth.

PITLESS ADAPTER: A special pipe fitting that fits on a well casing, below ground. It allows the pipe to pass horizontally through the casing so that no pipe is exposed above ground where it could freeze. The pump may be installed and removed without further need to dig around the casing. This is done by using a 1 inch threaded pipe as a handle.

SAFETY ROPE: Plastic rope used to secure the

pump in case of pipe breakage.

SUBMERSIBLE CABLE: Electrical cable designed for in-well submersion. Conductor sizing is specified in millimeters, or (in USA) by American Wire Gauge (**AWG**) in which a higher number indicates smaller wire. It is connected to a pump by a cable splice.

WELL SEAL: Top plate of well casing that provides a sanitary seal and support for the drop pipe and pump. Alternative: See *Pitless Adapter.*

Copyright ©1996 by Dankoff Solar Products, Inc.

APPENDIX C
Names and Numbers

These are the names of Solar Distributors in alphabetical order:

Alternative Energy:
    Order: (800) 777-6609
    Tech. Support: (800) 800-0246

Backwoods Distributors - (208) 263-4290

Dankoff Solar Pumping: (505) 473-3800

Direct Power Distributors: (800) 260-3792

Jade Mountain:
    Order: (800) 442-1972
    Tech. Support: (303) 449-6601

Positive Energy: (505) 424-1112

Real Goods:
    Order: (800) 919-2400

Staber Washers: (800) 848-6200

Trace Engineering: (360) 435-8826

APPENDIX D

WIRE SIZING CHART for 12 Volt based on 3% voltage drop

Use Chart to find amps on left column and wire size on top row. Find distance in the center rows and columns.
Example: If you want to run **10 amps** for **29 feet** you need a # 10 wire.

Chart is based on 12 Volt and a voltage drop of 3%.

If you want to calculate other distances, amps, and/or volt. drop %, you can use the following formula:

Distance X 12.9 X Amps
Volts X Volt.drop in %

$$\frac{175 \cdot 12.9 \cdot 13}{120 \cdot 3} = \frac{29347}{3.6} = 8151$$

The results are Circular Mill. which you can look up in the back of any NEC code book or the enclosed chart. It will give you the wire size. The value 12.9 is called the "K" value and is a combination of the Circular Mill and the resistively of the wire or R-Value which you also can find in the same chart at the back of any NEC code book. Since the values for stranded wire are all between 12.86 and 12.93 using 12.9 is a good average. This is the most accurate way of determining the correct wire size. Most charts published I found inaccurate and sometimes off by as much as 2 wire sizes, which can translate into a lot of money.
Here is one example how to calculate the wire size:

100 feet x 12.9 x 11 amps
---------------------------------- = 39,416 Circ. Mill
12 volt x 3% =(0.36)

Looking up the value in the enclosed chart will get you close to a # 4 wire.

WIRE SIZING CHART for 12 Volts, based on 3 % Voltage Drop

Wire size	14	12	10	8	6	4	2	1	1 O/D	2 O/D	4 O/D
Amps											
2	55	85	130	220	360	560	900	1200	1500	1900	2900
5	20	35	55	90	115	225	362	450	600	750	1200
10	12	18	29	45	57	112	180	230	300	380	590
15	7	11	18	30	47	75	120	155	200	250	400
20	6	8	13	22	36	56	90	120	190	190	300
25	--	6	11	17	29	45	72	100	120	150	235
30	--	--	6	15	25	37	60	75	100	125	200
50	--	--	--	6	15	22	36	45	60	75	120
100	--	--	--	--	7	11	18	22	30	37	60

Size AWG/kcmil	Area Cir. Mills
18	1620
18	1620
16	2580
16	2580
14	4110
14	4110
12	6530
12	6530
10	10380
10	10380
8	16510
8	16510
6	26240
4	41740
3	52620
2	66360
1	83690
1/0	105600
2/0	133100
3/0	167800
4/0	211600

APPENDIX E
WORKSHEET

WORK SHEET FOR A PHOTOVOLTAIC SYSTEM

Name of Load	Qua	Watts Ea	WH Day	Hrs. used	Total / Day	Typical W. per Item
LIGHTING						8 W to 20 W each
STEREO SYSTEM						Read Name plates
KITCHEN APPL.						Read Name Plates
ELECTR. TOOLS						Read Name Plates
TV						60W to 90W
WASHER						(Staber) 500W
DRYER						600W
CEILING FANS						30W
COMPUTER (Laptop)						50W
COMPUTER (Dtop.)						130W to 250W (incl. printer)
PRESSURE PUMP						DC: 200W / AC: 1000W
WELL PUMP						DC: 150W / AC: 1500W
VACUUM CLEANER						1000W to 1500W
OTHER LOADS						
TOTAL LOAD						Watt-Hours per Day
Add 20% for Losses						
TOTAL LOAD TO BE GENERATED						Watt-Hours per Day
ARRAY SIZE						
Divide total WHs. by Hrs. of average sunlight in your area						Watts to be recharged
Divide by Watts of Panel to be used						Number of Panels
Round off to # of Panels as needed by System Volts (12V, 24)						**Actual # of Panels**

BATTERY BANK			
Divide total WHs. by System Voltage (e.g. 4800 / 12V = 400 AH)	Amp Hours		
Add 30% for max discharge rate and losses	Amp Hours		
Multiply by days of reserve desired (3 to 5 days)	**Total AHs needed**		
Divide by AHs. of each Batt. Unit (e.g. Two 6V Batt. at 220 AHs from One 12V Unit at 220 AHs.	Units needed		
Multiply total Units by their number of Batteries	**Total Batts. needed**		

APPENDIX F
Average Hours of Sunlight Across U.S.

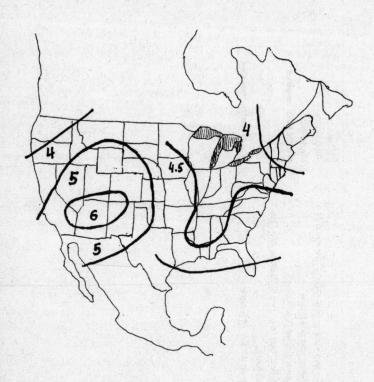

INDEX

211